PRAISE FOR **MURRAY LONGE**

"Murray distils decades of experience and complex concepts into an accessible roadmap."

KERI A. CLARKE - Start-up industry professional

"Murray has a deep understanding of what he is discussing."

KROL- Amazon Reviewer

PRAISE FOR **MURRAY LONGE'S OTHER BOOKS**

"Like sitting down for an in-depth chat with a knowledgeable, experienced friend in the business. Practical advice based on real world practical success. The writing style makes it easy to read and absorb what might otherwise be quite complex ideas."

WORDSMITH - Amazon Reviewer

"A concise and practical guide for entrepreneurs, packed with clear insights, case studies, and actionable steps. An invaluable resource."

KERI A. CLARKE

"This book is a very well-written, thorough look at every aspect of starting a business. I highly recommend it, especially as it's written by a successful entrepreneur, someone who has actually been through the start-up process from start to finish."

PETER G. - Amazon Reviewer

The Startup Success Series Book 2

Beyond The Minimum Viable Product

Build Products Customers Really Want

Murray Longe

WHITMAN
CROSS

Published by Whitman Cross Publishing, Melbourne, Australia.

First edition 2024

ISBN 978-1-7645437-8-1 (ebook)
ISBN 978-1-7645437-9-8 (paperback)

NATIONAL LIBRARY OF AUSTRALIA

A catalogue record for this book is available from the National Library of Australia

Disclaimer

The information in this book represents the author's views as of the date of publication. It is for general informational purposes only and is not intended as professional, legal, financial, or business advice. Because circumstances and regulations change, the author reserves the right to revise his views in light of new conditions. Readers should consider their own personal circumstances and consult with qualified professionals before taking any action based on the information in this book. While every effort has been made to ensure the accuracy and completeness of the information, the author and publisher assume no responsibility for any errors, omissions, or inaccuracies. Use of the information is at the reader's own risk.

For more information, visit whitmancross.com

WHITMAN
CROSS

Also by Murray Longe

How To Start Your Start-Up:
The 10 Key Steps To Get From Idea To Investment

Project Management for Startups:
Simple Tools and Templates to Launch, Lead, and Thrive

The Startup Success Series

#1 *Find Problems Worth Solving:*
How to Validate Your Startup Idea

#2 *Beyond The Minimum Viable Product:*
Build Products Customers Really Want

#3 *Scale Without Breaking:*
How to Grow Your Startup Without Losing What Made It Work

#4 *Beyond The Pitch Deck:*
How to Raise Startup Capital

Search All "Murray Longe"

in your amazon store

To the founders who choose clarity over complexity,
who understand that simple isn't always easy,
and who know that imperfect action beats perfect planning.

Your willingness to build, test, and learn
is what turns ideas into reality.

Murray Longe
August 2025

Table of Contents

Table of Contents

Table of Contents

Table of Contents

The Evolution Beyond MVP

Here's a question that reveals everything wrong with modern product development: "How long should an MVP take to build?"

I see this question constantly in startup forums, and the answers range from "two weeks" to "six months." But here's the thing: they're all missing the point. The real question isn't about timeline. It's about whether you understand what you're building and have chosen the right approach for your specific situation.

I've watched teams spend eight months building what they call an "MVP" that's actually a feature-complete platform with integrations, analytics, and mobile apps. And I've seen others ship genuinely minimal products in markets that expect polish, then act surprised when users ignore them. Both groups are failing, just in opposite directions.

This confusion isn't academic. It's expensive. Choose the wrong approach, and you'll either over-build for your resources or under-deliver for your market. Either mistake can kill your startup.

The MVP Trap

For over a decade, the startup world has treated MVP (Minimum Viable Product) as the One True Way to build products. Build fast, ship early, iterate based on feedback. It became startup gospel, repeated so often that questioning it feels like heresy.

But here's what nobody talks about: MVP was designed for a specific time and specific circumstances. When Eric Ries popularized it in "The Lean Startup," getting any working software to market was expensive and time-consuming. The primary competitive advantage was speed. If you could ship something functional while competitors were still planning, you won.

That world no longer exists.

Today, anyone can build a functional app in a weekend using AI tools and no-code platforms. The democratization of building means being first to market is just the entry fee. The real competition is for attention, trust, and emotional attachment. And that requires fundamentally different strategies.

Meanwhile, user expectations have skyrocketed. Today's customers have been trained by Apple, Google, and Netflix to expect polished experiences from day one. They have infinite alternatives and zero patience for products that feel unfinished.

In this new reality, "minimum" often means "forgettable," and "viable" often means "barely usable." Speed to market means nothing if nobody wants to stay.

The False Choice

The startup world presents MVP as if it's your only option: build fast and minimal, or spend years over-engineering. But this is a false choice that ignores how successful products actually get built.

Look at the companies dominating their markets today. They didn't win by shipping the fastest or the most minimal products. They won by creating experiences that users genuinely preferred over alternatives. Sometimes that meant building more than a traditional MVP. Sometimes it meant building less but with obsessive attention to user experience.

The key insight is this: there's no universally "right" approach to product development. The right approach depends on your market, your resources, your competitive landscape, and your specific goals.

Beyond the MVP Orthodoxy

This book introduces you to four distinct approaches to building products that people actually want:

MVP (Minimum Viable Product): Optimized for learning speed and assumption validation. Perfect for testing new market categories or when you have extreme resource constraints.

SLC (Simple, Lovable, Complete): Optimized for creating complete experiences that users love from day one. Ideal when you understand your market and want to build sustainable competitive advantages.

MLP (Minimum Lovable Product): Optimized for emotional connection and user advocacy. Essential in crowded markets where functional differentiation isn't enough.

MMP (Minimum Marketable Product): Optimized for revenue generation and sales success. Critical when you need to generate cash flow quickly or compete against established alternatives.

Each approach makes different trade-offs between speed, completeness, emotional engagement, and market readiness. The magic happens when you choose the approach that matches your specific situation and execute it with discipline.

What You'll Learn

This book will teach you how to:

- **Choose the right approach** for your market, resources, and goals using clear decision frameworks.
- **Execute each approach effectively** with specific strategies, team structures, and resource allocation plans.
- **Test and validate** your chosen approach using appropriate metrics and feedback methods.
- **Evolve your approach** as your product and market mature, including common transition patterns.
- **Avoid expensive mistakes** by understanding the hidden costs and requirements of each methodology.

More importantly, you'll learn to think strategically about product development rather than just following the latest methodology trends.

Building on Customer Discovery

This book builds directly on the foundation established in "Find Problems Worth Solving." That first book helped you identify problems worth solving and validate market demand. This book helps you solve those problems in ways that create sustainable businesses.

The connection is crucial: finding real problems is only half the equation. The other half is solving them in ways that create genuine value, emotional connection, and competitive advantage. That's where most startups fail, not in identifying problems but in building solutions people actually choose.

Your Path Forward

By the end of this book, you'll have frameworks for choosing and executing the right product development approach for your specific situation. You'll understand when to be minimal and when to be complete, when to optimize for learning and when to optimize for love.

But here's what I won't do: I won't prescribe a one-size-fits-all solution. The startup graveyard is full of companies that followed popular methodologies without considering whether those methodologies matched their reality.

Instead, I'll give you the tools to make intelligent decisions based on your unique circumstances. Because the best product development approach isn't the one that worked for some famous company. It's the one that works for your market, your team, and your goals.

Your customers are waiting for products that not only solve their problems but also solve them in ways that feel delightful, complete, and worth recommending. Your market is ready for solutions that stand out from the sea of minimal, barely-viable alternatives.

The question isn't whether you can build something people want. The question is whether you'll choose the *right way* to build it.

Let's find out.

1
Rethinking the Minimum Viable Product

Let me start with a confession. I used to be a true believer in MVP orthodoxy. Build fast, ship early, iterate quickly. I preached lean startup principles with the fervor of someone who had just discovered fire. It seemed like the perfect antidote to the software industry's history of bloated products that took years to develop and often missed the mark entirely.

But after watching hundreds of startups follow the MVP playbook religiously only to fail spectacularly, I've come to a different conclusion. It's like watching people follow a recipe for chocolate cake and consistently producing hockey pucks. The MVP approach isn't wrong, but it's dangerously incomplete. And in today's market, incomplete strategies lead to complete failures.

The Original MVP Promise

When Eric Ries introduced the Minimum Viable Product concept in "The Lean Startup," he was solving a real problem. Software companies were spending years building products in isolation, only to discover that customers didn't want what they'd built. The MVP was meant to be a learning vehicle: the smallest thing you could create to test your core assumptions about customer behavior.

The original definition was clear: "A minimum viable product is that version of a new product which allows a team to collect the maximum amount of validated learning about customers with the least effort."

Notice what's missing from that definition? Any mention of launching a business, generating revenue, or creating customer satisfaction. The MVP was designed to facilitate learning, not to be a product that customers would love.

Somewhere along the way, that nuance got lost. It's like making a photocopy of a photocopy, where each generation gets progressively worse quality. "Build something to learn" became "build something fast" became "build something cheap" became "build something mediocre and hope for the best."

Where MVP Thinking Goes Wrong

The problem isn't with the MVP concept itself. It's with how founders interpret and implement it. Three critical misconceptions have turned a useful learning tool into a recipe for mediocrity:

Misconception 1: Minimum Equals Fastest

Founders hear "minimum" and think "quick and cheap." They strip away everything that isn't absolutely essential, believing that speed to market trumps all other considerations. It's the entrepreneurial equivalent of showing up to a dinner party with gas station flowers and wondering why nobody's impressed. But minimum doesn't mean rushed. It means intentionally focused.

The difference is profound. A rushed product cuts corners and delivers a poor experience. A focused product deliberately chooses what to excel at and what to exclude.

Instagram's origin story is a lesson in focus. The founders didn't build a bloated all-purpose app; they stripped Burbn down to photo sharing and filters, creating a simple product that made ordinary mobile photos look visually appealing from day one and quickly resonated with users. Video, messaging, and other social features came later.

The lesson here runs deeper than feature selection. Instagram's founders understood that customers don't judge products by what they lack. They judge them by how well they deliver on their core promise. A product that does one thing brilliantly will always outperform a product that does five things adequately.

This distinction becomes critical when you consider switching costs. Customers aren't just evaluating your product in isolation. They're comparing it to their current solution, which might be imperfect but is familiar. Your product needs to be significantly better to justify the effort of switching. "Faster to build" doesn't translate to "easier to adopt."

Misconception 2: Viable Means "Good Enough"

This is the killer. And I mean that literally. This misconception has killed more startups than overconfident projections and underestimated timelines combined. Founders interpret "viable" as "functional enough that people won't immediately delete it." But in a world where users have infinite alternatives, "good enough" is a death sentence.

Your customers don't want viable. They want valuable. They don't want functional. They want delightful. The gap between those expectations and typical MVP execution is where startups go to die.

Color Labs raised $41 million and built an MVP app for photo sharing that was technically viable. It had the core features, it solved a real problem, and it was technically fine. It worked. But it never became something people really cared about. It wasn't better than existing solutions in any meaningful way. Viable but forgettable. Users tried it, then moved on to Facebook and Instagram because those products were more compelling. It's a bit like opening a restaurant that serves food that won't poison you, then wondering why people keep going to the place down the street that serves food they actually enjoy.

The company's failure after about 20 months shows the real lesson: in a crowded market, it's not enough to build something functional. You have to build something people actually want.

They missed product-market fit. They built something that worked instead of something that mattered. In a competitive market, working isn't enough. You need to forge an emotional connection that makes users choose you over perfectly adequate alternatives.

This emotional dimension is where most MVPs fail. They solve problems logically but ignore the human elements that drive adoption: pride, joy, confidence, and belonging. A truly successful product makes users feel something positive about themselves when they use it.

Misconception 3: Learning Equals Revenue

The most dangerous assumption is that an MVP that validates learning automatically validates a business model. But customers who are willing to try your product for free, or even provide feedback, are fundamentally different from customers who are willing to pay for it. It's the difference between someone who'll taste your cooking and someone who'll pay for dinner at your restaurant.

This learning-revenue gap explains why many "successful" MVPs never become successful businesses. These products might show high engagement, positive feedback, and rapid user growth, but when the time comes to convert users to paying customers, the conversion rates are dismal. The founders had optimized for learning, not buying.

The problem stems from different user motivations. People who test products are often motivated by curiosity, a desire for novelty, or a desire to contribute to the development of new products. People who buy products are motivated by urgent problems, clear value propositions, and confidence in outcomes. These are fundamentally different mindsets that require different validation approaches.

This disconnect becomes particularly dangerous when startups use engagement metrics as proxies for purchase intent. High daily active user counts can mask low willingness to pay. Strong feature usage can hide weak value perception. Positive feedback can obscure price sensitivity. Without directly testing purchase behavior, learning validation becomes misleading validation.

The Three Fatal Flaws of MVP-First Thinking

When you dig deeper into why MVP-first approaches fail, three patterns emerge:

Fatal Flaw 1: The Feature Fallacy

MVP thinking encourages founders to think in terms of features: what's the minimum set of features needed to be viable? This leads to products that feel like feature lists rather than coherent solutions.

Users don't buy features. They buy outcomes. They don't want a tool with basic project management capabilities. They want to feel in control of their work. They don't want an app with messaging functionality. They want to feel connected to their team.

When you start with "What's the minimum?" you optimize for less rather than better. When you start with "What outcome are we creating?" you optimize for value.

This feature-centric thinking creates products that feel mechanical rather than magical. Each feature exists because it passed some functional test, not because it contributes to a delightful user experience. The result is products that check boxes without touching hearts. It's like designing a car by listing all the mechanical components it needs, then wondering why nobody falls in love with driving it.

Consider the difference between two approaches to building a fitness app:

- The **feature**-fallacy approach asks: "What's the minimum we need? Exercise tracking, basic reporting, social sharing."
- The **outcome**-focused approach asks: "How do we make people feel successful about their health?"

This leads to entirely different design decisions, from encouraging micro-habits to celebrating small wins.

Fatal Flaw 2: The Comparison Trap

MVP logic often sounds like this: "Competitor X has 50 features, so our MVP with 5 features will be faster to build and launch." But customers aren't comparing your 5 features to their 50 features. They're comparing your 'entire experience' to their 'entire experience'. It's like entering a swimming competition with perfect form for the backstroke, then wondering why you lost the freestyle race.

If Competitor X solves their problem completely (even if imperfectly), and your MVP only solves part of their problem, you lose. It doesn't matter that you're faster or leaner. Incomplete solutions rarely beat complete ones, even when the complete solution is flawed.

Slack didn't beat email by being a minimal version of email. They won by providing a complete solution to team communication, something email has never attempted to be.

The comparison trap becomes more dangerous when you consider that customers don't make rational feature-by-feature comparisons. They make emotional, holistic judgments based on their complete experience. A product that does everything adequately often beats a product that does some things brilliantly but leaves gaps in the user journey.

This is why many MVP products feel like demos rather than products. They showcase potential rather than delivering value. Customers can see what the product might become, but they can't get their work done with what it is today. In competitive markets, potential isn't enough.

Fatal Flaw 3: The Patience Paradox

MVP advocates argue that launching early allows you to iterate based on user feedback. But this assumes your early users will have the patience to stick around while you figure things out.

In reality, most users give you one chance to make a first impression.

If your MVP disappoints them, they don't bookmark it to check back later. They forget about you and move on. By the time you've iterated your way to a great product, your window of opportunity has closed. It's like getting one chance to make a first impression at a networking event, then spending the rest of the night explaining to empty chairs how much better your pitch has gotten.

This patience paradox is particularly brutal in consumer markets where switching costs are low and alternatives are abundant. Enterprise users might stick with an improving product because switching has high costs. But consumer users can try a dozen alternatives in the time it takes you to ship your next iteration.

The cost of this impatience extends beyond individual users. In our connected world, first impressions spread rapidly through social networks and review platforms. A disappointing launch not only loses individual users but also creates negative social proof, making future customer acquisition more expensive.

Smart founders recognize that you get one chance to launch in each market segment. You can iterate privately, but your public launch needs to deliver a complete, lovable experience from day one.

The MVP Reality Gap

Here's what the MVP Reality Gap looks like in practice:

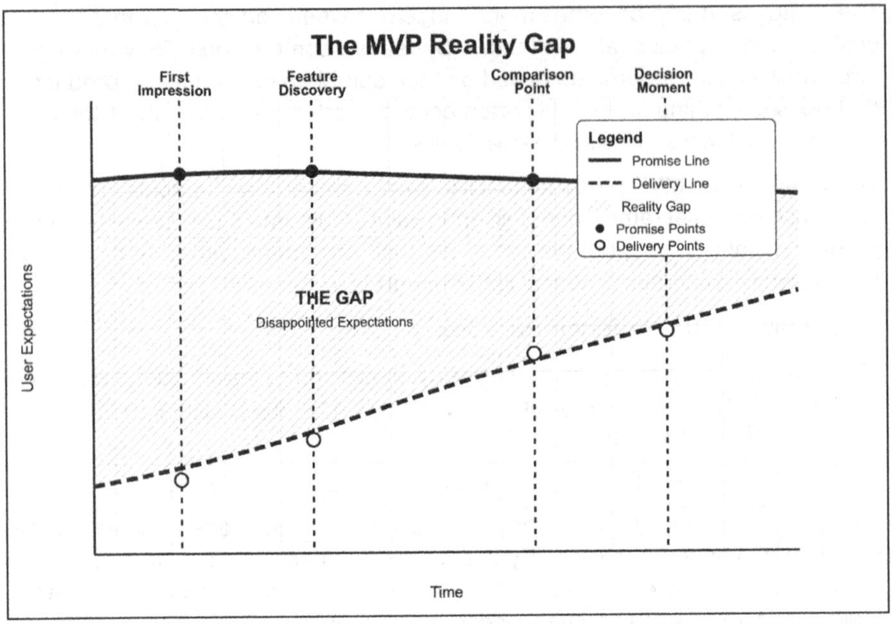

The MVP Reality Graph has two lines diverging over time:

- **The Promise Line** starts high, representing the complete solution implied by your marketing, pitch, and vision.
- **The Delivery Line** starts low, representing the minimal functionality of your actual MVP.
- **The Gap** between these lines represents disappointed expectations.

Along the timeline, you see critical moments:

- **First Impression**: Where the gap is widest, and most users decide whether to continue.
- **Feature Discovery**: Where users realize limitations of the MVP.
- **Comparison Point**: Where users evaluate alternatives and see what's missing.
- **Decision Moment**: Where users make their final choice to stay or leave.

The wider the gap between promise and delivery, the higher the abandonment rate at each critical moment. Successful products minimize this gap from day one.

Why Customers Hate MVPs

Let's be honest about what most MVPs actually deliver to customers:

1. **The Broken Promise Experience**

   MVPs often promise more than they deliver. The marketing claims to solve a complete problem, but the product only addresses part of it. Customers feel misled, not delighted. It's like ordering a "complete breakfast" and getting a single piece of toast. Technically, toast is breakfast food, but good luck explaining that to your hungry stomach.

   This Broken Promise Experience is particularly damaging because it destroys trust before you've had a chance to build it. Customers approach new products with skepticism, and when your MVP confirms their worst expectations about startup reliability, you've lost them permanently.

   The promise-delivery gap also affects word-of-mouth marketing. Customers who feel misled don't just leave silently. They warn others about their disappointing experience. In markets where reputation matters, this negative word-of-mouth can be fatal.

2. **The Beta Tester Burden**

   MVPs essentially ask customers to do quality assurance for free. They're expected to encounter bugs, work around missing features, and provide feedback while paying for the privilege. Most customers didn't sign up to be unpaid consultants. It's like inviting someone to dinner and then asking them to cook, serve, and clean up while critiquing your recipe.

   Professional beta testers and early adopters might accept this burden in exchange for early access or influence over product direction. But mainstream customers have different expectations. They want products that work reliably without requiring their input or patience.

   The Beta Tester Burden becomes particularly problematic when your product is integrated into customers' critical workflows. If your MVP causes them to miss deadlines, lose data, or appear unprofessional to their clients, the cost of testing your product far exceeds any benefit you provide.

3. **The Feature Hostage Situation**

   MVP products often feel held together with digital duct tape. Customers want to accomplish something simple, but they hit limitations, workarounds, and "Coming Soon" messages at every turn. The product becomes a source of frustration rather than a source of value.

   This Feature Hostage Situation creates a particularly toxic dynamic. Customers start using your product because it promises to solve their problem, but then they discover they can't complete their workflow without missing features. Now they're stuck. They've invested time learning your product, but they can't get their work done.

Smart competitors exploit this hostage situation by building complete solutions that let customers accomplish their goals without workarounds. They win customers not through innovation, but through competence. It's the business equivalent of being the friend who actually shows up on time while everyone else makes excuses.

4. The Comparison Disadvantage

When customers compare your minimal product to established alternatives, you lose on almost every dimension except "newness." That's not enough to drive switching behavior.

This Comparison Disadvantage becomes more severe over time. While you're iterating your way to completeness, established competitors are also improving their products. The gap you're trying to close keeps moving. What seems like a 6-month development timeline becomes an 18-month race you can never win.

The only way to overcome the Comparison Disadvantage is to be dramatically better in your core strength area. But MVP thinking often prevents this dramatic differentiation by encouraging breadth over depth.

When MVP Works

I'm not completely anti-MVP. That would be like being anti-hammer just because people keep hitting their thumbs. There are specific situations where the traditional MVP approach makes perfect sense:

- **When You're Testing a New Market Category**

  If you're creating something truly novel, a product category that doesn't exist, then MVP makes sense. Customers can't compare you to established alternatives because there aren't any. Your job is to validate that the category itself has merit.

  Uber's early MVP made sense because ride-sharing didn't exist. They weren't competing against better ride-sharing apps. They were validating whether people would summon rides with strangers via their phones.

  In these situations, the learning value of MVP outweighs the experience limitations because customers have no baseline for comparison. You're not trying to be better than existing solutions; you're trying to prove that any solution is needed.

  However, even in new categories, you must deliver a comprehensive experience within your chosen scope. Uber didn't build a partial transportation service. They developed a comprehensive and reliable method for travelling from point A to point B, which happened to utilise a new model.

- **When Your Constraints Are Extreme**

  If you have severe resource constraints (limited time, money, or team), then MVP might be your only viable option. Better to test core assumptions with something minimal than to never test at all.

  But recognize this as a constraint-driven decision, not an optimal strategy. Plan your path from MVP to something customers will love. Set clear milestones for when you'll invest in completeness and emotional design.

  The danger of constraint-driven MVPs is that they can become permanent. Teams get comfortable with minimal and lose sight of the need to evolve toward lovable. It's like wearing a temporary cast for so long that you forget your arm is supposed to bend. Set explicit triggers for when you'll graduate beyond MVP thinking.

- **When You're Targeting Early Adopters**

  Some customer segments, particularly technical users or industry insiders, are willing to tolerate minimal products in exchange for early access to innovation. They understand they're trading convenience for the opportunity to be first.

  These customers often derive value from the novelty itself. Being first to use new technology can provide social status, competitive advantage, or personal satisfaction that outweighs functional limitations.

  But don't confuse early adopter tolerance with mainstream market acceptance. What works for developers often fails with consumers. Plan your evolution from early adopter product to mainstream product carefully.

- **When Speed Is Your Only Competitive Advantage**

  If your market window is extremely narrow, and being first matters more than being good, MVP can work. But this is increasingly rare in today's market, where being second or third with a superior product often beats being first with an inferior one.

  Even in time-sensitive markets, remember that speed to market isn't the same as speed to delivering customer satisfaction. Sometimes it's better to launch second with a complete solution than first with an incomplete one.

The Real MVP Success Stories

Let's examine some companies that are often cited as MVP success stories and see what they actually built:

Dropbox: The Video That Wasn't Minimal

Drew Houston's famous Dropbox video is often hailed as the perfect MVP. But look closer. That video wasn't minimal; it showed a flawless user experience, seamless file syncing, and effortless sharing—promising a complete solution to a complete problem.

The genius wasn't in building something minimal. It was in validating complete value without building anything at all.

The video succeeded because it let viewers envision using the product in their daily workflow without frustrations. This completeness of vision, even pre-product, sparked genuine excitement and proved real demand.

Airbnb: The Manual MVP That Delivered Excellence

Airbnb's founders didn't build a minimal platform. They *manually* delivered an excellent experience. They photographed apartments professionally, handled payment processing personally, and ensured every guest had a great stay. The experience was complete and lovable from day one.

The platform was minimal, but the customer experience was maximal.

This distinction is crucial. Airbnb validated their business model by delivering complete value manually before building technology to scale that value. They proved that strangers would stay in each other's homes, but only when the experience was carefully curated and professionally managed.

Zappos: Proving the Complete Value Proposition

Nick Swinmurn didn't build a minimal e-commerce site. He proved that the complete value proposition (selection, service, and satisfaction) could work for online shoe sales. He manually sourced inventory and provided exceptional service to prove that customers would buy shoes online if the experience was complete.

The Zappos validation focused on the complete customer journey, from browsing to buying to returning. They didn't test whether people would visit a shoe website. They tested whether people would complete purchases and remain satisfied with the entire experience.

The Path Forward: Beyond Minimum

If traditional MVP thinking is flawed, what's the alternative? The answer isn't to abandon fast iteration or customer validation. It's to change what you're optimizing for.

Instead of asking "What's the minimum we can build?" ask these questions:

1. **What's the simplest complete solution to this problem?** Focus on completeness within a narrow scope rather than partial solutions to broad problems.
2. **What would make customers love this from the first interaction?** Optimize for emotional engagement, not just functional utility.
3. **What's the smallest thing we can build that people will pay for?** Shift from learning validation to revenue validation.
4. **How can we deliver maximum value with focused scope?** Choose what to excel at, not what to exclude.

This isn't about building more features or taking longer to launch. It's changing your definition of success from "**shipped something**" to "**created value.**"

The New Product Development Hierarchy

Instead of the traditional MVP approach, consider this hierarchy:

1. **Problem Validation** (from Book 1): Confirm the problem is real and painful.
2. **Solution Validation**: Test that your approach can solve the problem completely.
3. **Value Validation**: Verify that customers are willing to pay for your solution.
4. **Experience Validation**: Ensure the complete user journey is delightful.
5. **Scale Validation**: Prove the model works at increasing volumes.

Notice that building comes after validating. And when you do build, you build to make your customer love your product, not just viability.

This hierarchy ensures that each validation step builds on the previous one. You don't test solutions until you've confirmed problems. You don't test pricing until you've validated solutions. You don't optimize experiences until you've confirmed value. This staged approach prevents the common mistake of optimizing the wrong things.

Measuring Success Beyond Viability

If you're moving beyond MVP thinking, to measure your success, you need metrics that go beyond traditional MVP measures:

Traditional MVP Metrics:

- User signups
- Feature usage
- Customer interviews completed

- Iteration velocity

Beyond-MVP Metrics:

- Net Promoter Score (NPS)
- Customer willingness to pay
- Organic referral rates
- Feature completion rates
- Customer lifetime value
- Time to core value achievement
- Support ticket volume per user
- Retention at 30, 60, 90 days

The goal isn't just to validate that people will use your product. It's to validate that they'll love it, pay for it, and recommend it.

These advanced metrics require more sophisticated tracking but provide much better predictors of long-term success. A product with high NPS and strong organic referrals has a much better chance of scaling than a product with high signup rates but poor retention.

Making the Mental Shift

Moving beyond MVP thinking requires a fundamental mental shift:

- From **"What's the least we can do?"**
  to **"What's the most value we can create?"**
- From **"How fast can we ship?"**
  to **"How quickly can we create customers who love us?"**
- From **"What features do we need?"**
  to **"What experience are we creating?"**
- From **"Is this viable?"**
  to **"Is this valuable?"**

This shift doesn't mean abandoning speed or efficiency. It means optimizing for *customer value* rather than *development convenience*.

The mental shift also affects how you think about competition. Instead of trying to match competitors feature-for-feature, you focus on delivering superior outcomes in your chosen area. Instead of rushing to market with partial solutions, you strive for customer satisfaction with complete solutions.

Your MVP Audit

If you're currently building or planning an MVP, ask yourself these hard questions (and try not to rationalize the uncomfortable answers):

1. **The Honesty Test**:
   If your MVP launched today, would customers choose it over existing alternatives? If not, why not?
2. **The Payment Test**:
   Would customers pay your target price for your MVP? If you're not sure, you're not ready to build.
3. **The Recommendation Test**:
   Would customers recommend your MVP to friends? If it's not good enough to recommend, it's not good enough to launch.
4. **The Completeness Test**:
   Does your MVP solve the customer's problem completely, or does it create new problems they need to solve elsewhere?
5. **The Love Test**:
   Will customers feel delighted by their first interaction with your product, or just satisfied that it works?
6. **The Workflow Test**:
   Can customers complete their entire workflow using your MVP, or do they need to use other tools to finish their tasks?
7. **The Comparison Test**:
   When customers compare your MVP to their current solution, what specific advantages will make them switch?
8. **The Patience Test**:
   If customers try your MVP and find limitations, will they wait for improvements or look for alternatives?

If you can't answer these questions confidently, you might be falling into the MVP trap.

Preparing for the Alternative

The next chapter will introduce you to the Simple, Lovable, Complete (SLC) framework. This approach maintains the speed and learning benefits of MVP while creating products that customers want to use, are willing to pay for, and recommend.

But first, take an honest look at your current approach. Are you building something minimal or something meaningful? Because in today's market, there's a world of difference between the two. And trust me, your customers can tell the difference faster than you can say "lean startup methodology."

The companies that succeed in the post-MVP world aren't the ones that build fastest. They're the ones that create the strongest emotional connections with customers from Day One. They understand that in a world where anyone can build a product over a weekend (thanks, Vibe Coding!), the winners are those who create products worth loving.

2

Simple, Lovable, Complete - The SLC Revolution

A few years ago, I was reviewing product pitches for an early-stage accelerator when one presentation perfectly captured everything wrong with modern product development. A founder excitedly explained his MVP: "It's like Slack, but for construction teams, but minimal."

I asked a simple question: "Would you use this to manage your own construction project?"

Long pause. "Well, not yet. It's missing a few features right now. But once we iterate..."

I closed my laptop. "Come back when you'd use it for your own project."

That conversation stayed with me because it exposed the core problem with MVP thinking. We've become so obsessed with building the bare minimum that we've forgotten to create something anyone would want to use.

Enter Simple, Lovable, Complete (SLC), a framework that shifts how product development is approached.

What Is Simple, Lovable, Complete?

The SLC framework was popularized by Jason Cohen, founder of WP Engine, who grew frustrated watching startups build products that were technically viable but emotionally vacant.

Instead of asking "What's the minimum we can get away with?" SLC asks, "What's the simplest *complete* solution that people will love?"

Let me break down each component:

Simple:
The product should be straightforward and intuitive. But simple doesn't mean minimal. It means thoughtfully designed for clarity and ease of use. Think of an

iPhone. It's simple to use, but it's certainly not minimal in its capabilities or construction quality.

Lovable:
The product should create an emotional connection from the first interaction. Users should feel delight, satisfaction, or pride when using it. Lovable products don't just solve problems. They make people feel good about themselves while solving those problems.

Complete:
The product should solve the customer's problem fully within its chosen scope. No "Coming Soon" messages, no workarounds, no forcing users to cobble together multiple tools to get their job done.

The magic happens when these three elements all work together. A product that's simple but not complete feels like a demo. A product that's complete but not simple feels overwhelming. A product that's simple and complete but not lovable feels boring.

You need all three.

Simple vs. Minimum: The Critical Distinction

This is where most founders get confused. They think “simple” and “minimum” are the same thing. They're not. They're often opposites.

Steve Jobs, the co-founder of Apple, once said,

> "Simplicity is the ultimate sophistication. It takes a lot of hard work to make something simple, to truly understand the underlying challenges and come up with elegant solutions."

Jobs understood that true simplicity comes from deep comprehension, not just a surface-level reduction.

This thinking aligns perfectly with SLC principles. Where MVP strips features to ship fast, SLC requires the hard work Jobs described: understanding your users so deeply that you can build something simple yet complete. The result is a product that feels effortless to use, precisely because of the effort invested in making it so.

Minimum asks: "What can we *remove*?" Simple asks: "What can we *clarify*?"

Minimum results in products that *do less*. Simple results in products that *feel easier*.

Let me illustrate with an example. Imagine you're building a calendar app. The minimum approach would be: "Users can create events and view them in a list." Technically viable, sure. But hardly compelling.

The simple approach would be: "Users can effortlessly schedule meetings that are planned to happen." This might require more features (calendar integration, email reminders, conflict detection), but the experience feels simpler because users don't have to think about the complexity.

Google understood this with Gmail. Their "simple" email interface in reality had more sophisticated features than competitors (conversation threading, powerful search, generous storage). But it felt simpler because they hid complexity behind intuitive design rather than simply removing features.

The key insight is that simple doesn't mean doing less. It means making the complex feel effortless. Sometimes that requires building more, not less.

Building for Customer Love from Day One

Here's something that might sound fluffy but is critical for business success: customers need to feel something positive when they use your product. Not just satisfaction that it works, but genuine delight in the experience.

This isn't touchy-feely startup nonsense. It's cold, hard economics. Businesses making products that people love have better customer retention, higher customer lifetime value, stronger word-of-mouth growth, and more pricing power. Customer love is a competitive moat.

But how do you build for love? It starts with understanding that love isn't an accident. It's the result of deliberate design decisions that prioritize human emotion alongside functional utility.

The Delight Stack

When it comes to delighting your customers with your product, think of "delight" as having three layers:

Layer 1: Functional Delight
The product does what it promises reliably and efficiently. This is table stakes. No one loves products that don't work. But only working isn't enough.

Layer 2: Usability Delight
The product is not just functional but *enjoyable* to use. The interface feels intuitive, the interactions feel natural, and the overall experience flows smoothly. This is where many MVP products fail. They achieve functional delight but ignore usability delight.

Layer 3: Emotional Delight
The product makes users feel good about themselves. It boosts their confidence, saves them from embarrassment, makes them look smart to their peers, or simply brings joy to mundane tasks. This is where love happens.

Consider Slack's early success. At the functional level, it sent messages (like email). At the usability level, it organized conversations better than email. But at the emotional level, it made teams feel more connected and productive. People didn't just use Slack; they evangelized it because it made them feel like they were part of something better.

The Love Multipliers

Certain design decisions amplify the emotional impact of products:

Anticipation:
Great products anticipate what users need before they ask for it. Spotify's Discover Weekly doesn't just play music; it makes users feel like someone who truly understands their taste has curated music just for them.

Recognition:
Lovable products make users feel seen and understood. When Notion automatically suggests a template based on what you're trying to build, it's not just helpful – it feels like the product "gets" you.

Progression:
People love feeling like they're improving or advancing. Duolingo doesn't just teach languages; it makes users feel proud of their streak and progress. The learning is functional, but the pride is emotional.

Surprise:
Thoughtful, unexpected moments create disproportionate delight. When Mailchimp celebrates you sending a campaign with a quirky animation, it turns a mundane business task into a small moment of joy.

The crucial insight is that these multipliers don't require massive feature development. They require empathy and attention to the emotional journey users take through your product.

Why Completeness Beats Minimalism

Here's where SLC fundamentally diverges from MVP thinking: completeness matters more than minimalism. Not *feature* completeness (throwing in every possible feature), but *solution* completeness (solving the customer's problem end-to-end within your chosen scope).

Think about it from the customer's perspective. They don't care about your development philosophy or resource constraints. They have a job to be done, and they want tools that help them complete that job without frustration or gaps.

The Job Completion Principle

Every customer interaction with your product is part of a larger job they're trying to complete. Your product doesn't exist in isolation. It's part of their workflow, their day, their goals.

If your product only handles part of their job, you're forcing them to context switch to other tools. Each context switch creates friction. Faced with enough friction, they'll find a tool that eliminates the switching entirely.

Let's say someone wants to create a simple website for their business. The MVP approach might be: "Here's a tool that lets you drag and drop elements onto a page." Technically, that enables website creation.

The SLC approach would be: "Here's a tool that takes you from idea to published website with your domain connected and looking professional." This might require more backend complexity (domain management, hosting, templates), but it solves the complete job.

This is why Squarespace succeeded while many simpler website builders failed. They didn't just let customers build websites. They handled hosting, domains, templates, mobile optimization, and even basic SEO. The complete job was done within their ecosystem.

The Integration Advantage

Complete solutions often win because they eliminate integration headaches. When everything works together seamlessly, the user experience becomes magical in a way that cobbled-together tools never achieve.

Consider how Notion approached note-taking and productivity. Instead of building a minimal note-taking app, they built a complete workspace that handled notes, databases, tasks, and collaboration. The integration between these elements creates emergent value that specialized tools can't match.

Users don't just get better note-taking; they get a thought-to-action workflow that feels cohesive and powerful. That integration is worth more than the sum of individual features.

Real-World SLC Success Stories

Let's examine some companies that embodied SLC principles and see how it affected their growth:

Google Docs: Simple, Lovable, Complete Document Creation

When Google launched Google Docs, Microsoft Word had been the dominant word processor for decades. Word was feature-rich, powerful, and entrenched. How could Google compete?

They didn't try to match Word feature-for-feature. Instead, they built a complete solution to collaborative document creation that was simple to use and delightful to share.

Simple: Clean interface with only essential formatting tools visible. No confusing ribbon interfaces or buried features.

Lovable: Real-time collaboration felt magical in 2006. Seeing someone else's cursor moving and text appearing live created genuine delight. Comments and suggestions made feedback feel conversational rather than formal.

Complete: Despite having fewer features than Word, Google Docs was complete for collaborative document creation. You could write, format, share, comment, and publish without leaving the tool or dealing with version conflicts.

The result? Google Docs didn't win by being a better word processor. It won by being a complete collaborative writing experience that felt effortless and delightful.

Snapchat: The Complete Ephemeral Sharing Experience

Snapchat is often misunderstood as a "simple" messaging app. But look closer, and you'll see a complete solution to casual photo sharing that was both simple to use and emotionally engaging.

Simple: Take a photo, add text or a drawing, and send to friends. The core interaction was intuitive for anyone who'd used a camera phone.

Lovable: The ephemeral nature made sharing feel playful rather than permanent. Filters and lenses added whimsy. The streak feature created affectionate habits between friends.

Complete: Snapchat didn't just send photos. It handled the complete casual sharing experience: capture, enhance, send, view, and automatic cleanup. No photo management, no storage worries, and no privacy concerns about permanent records.

This completeness is why Snapchat survived despite Facebook copying its core features. Facebook could copy disappearing messages, but they couldn't copy the complete cultural experience Snapchat had created around casual, ephemeral sharing.

Superhuman: Email as a Complete Productivity Experience

Superhuman dared to reinvent email—one of software's most entrenched categories—against giants like Gmail and Outlook.

Simple: Keyboard shortcuts for every action. A clean, focused interface. Email triage boiled down to fast decisions: archive, reply, or snooze.

Lovable: Blazing performance transformed sluggish inboxes into responsive tools. The onboarding turned users into email ninjas, with subtle animations reinforcing every swipe and command.

Complete: Superhuman went beyond basic sending and receiving. It owned the full productivity workflow—rapid triage, intelligent reminders, scheduled sends, read receipts, and contextual contact insights.

The lesson? Superhuman didn't just improve email. It redefined it as a seamless, delightful system that made professionals faster and happier. Users don't just switch email clients for slightly better features. But they will switch for a *complete* experience that makes them dramatically more effective at email management.

The SLC Implementation Playbook

Understanding SLC conceptually is one thing. Building SLC products is another. Here's a practical framework for implementing SLC thinking in your own product development:

Step 1: Define Your Job Scope

Before you design anything, get crystal clear on the complete job your customers are trying to do. Not just the functional task, but the emotional and social context around it.

Ask these questions:

- What triggers someone to need this solution?
- What's the ideal outcome they're hoping for?
- What happens before and after they use your product?
- What could go wrong that would make them feel frustrated or embarrassed?
- What would make them feel proud or successful?

For example, if you're building a presentation tool, the job isn't just "create slides." The complete job might be "confidently deliver a persuasive presentation that achieves my goal and makes me look professional."

This broader job definition reveals opportunities that pure slide creation tools miss: speaker notes, presenter confidence features, audience engagement tools, post-presentation follow-up.

Step 2: Map the Emotional Journey

Most product teams map user flows but ignore emotional flows. For each step in your user's journey, consider:

- How are they feeling when they start this step?
- What emotions does your current design create?
- What emotions would you ideally create?
- What specific design decisions could shift their emotional state?

Create an emotional journey map that looks something like this:

Step 1: Discovery

- Current emotion: Frustrated with existing solutions
- Ideal emotion: Hopeful that this might work better
- Design opportunity: Clear value proposition that acknowledges their frustration

Step 2: First Use

- Current emotion: Cautious, possibly overwhelmed
- Ideal emotion: Confident and quickly successful
- Design opportunity: Guided experience that creates early wins

Step 3: Regular Use

- Current emotion: Habitual, possibly bored
- Ideal emotion: Increasingly competent and proud
- Design opportunity: Progressive disclosure that reveals power over time

This emotional mapping often reveals design opportunities that functionality mapping misses.

Step 3: Design for Complete Success

Instead of designing features, design for complete success scenarios. Start with your user achieving their ideal outcome and work backward.

Create success scenarios like: "Sarah uses our tool to create a presentation that impresses her boss and leads to her promotion."

"Marcus uses our platform to launch a campaign that sells out his product and establishes his brand."

"Jennifer uses our service to plan an event that her friends still talk about months later."

These success scenarios force you to think beyond your product's boundaries. What would need to be true for Sarah's presentation to impress her boss? It's not just good slides; it's relevant content, confident delivery, professional design, and smooth technical execution.

This success-backward design often reveals missing pieces in your solution that you might have dismissed as "nice-to-have" features.

Step 4: Ruthlessly Simplify the Path

Once you know the complete job and desired emotions, ruthlessly eliminate everything that doesn't contribute to that success. But notice: you're not removing features for the sake of minimalism. You're eliminating complexity that gets in the way of success.

Ask for every element of your product:

- Does this help users achieve their complete success scenario?
- Does this create positive emotions or neutral/negative ones?
- If we removed this, would success become harder or just different?
- Could we achieve the same outcome with fewer steps or decisions?

This doesn't mean focusing on fewer features. It means having zero unnecessary friction between your user and their success.

Step 5: Build Love into the Details

Love lives in the details. Once you have the complete user journey mapped and simplified, look for moments to exceed expectations in small but meaningful ways.

These might include:

- Celebratory moments when users achieve milestones
- Helpful suggestions that anticipate needs
- Error messages that are encouraging rather than blaming
- Delightful animations that make interactions feel alive
- Personalization that makes users feel understood

The key is that these details should feel natural and purposeful, not gratuitous. They should reinforce the emotions you want users to feel about their success.

SLC vs MVP: A Side-by-Side Comparison

Let me illustrate the practical differences between SLC and MVP approaches with a concrete example. Imagine you're building a tool to help small businesses manage their social media presence:

	The MVP Approach	The SLC Approach
Core Features/Job:	Post scheduler, basic analytics	Help small business owners feel confident about their social media presence
Success Criteria:	Users can schedule posts and see basic metrics	Users consistently post engaging content and feel proud of their social media growth
Timeline:	6 weeks to build, launch quickly	12 weeks to build, launch when complete

User Experience:	"Here's a basic tool. We'll add more features based on your feedback."	"Here's everything you need to succeed on social media, designed to make you look professional and grow your audience."

The Different Outcomes:

MVP outcome: Users sign up, try the scheduler, realize it doesn't integrate with their preferred platforms, requires separate tools for content creation and performance tracking. Some stick around hoping for improvements, but most churn to more complete solutions.

SLC outcome: Users sign up and immediately see how the tool fits into their complete social media workflow. They feel confident using it because it handles content suggestion, optimal timing, cross-platform posting, engagement tracking, and performance insights in one cohesive experience.

The SLC product takes twice as long to build but creates users who are more likely to stick around, pay higher prices, and recommend the product to others.

The SLC vs MVP User Experience Journey

The User Experience Journey (shown on the next page) illustrates how an MVP creates an emotional rollercoaster that often ends in disappointment, while an SLC model creates a steady climb toward satisfaction and advocacy. Key decision points are marked where users either commit deeper or start looking for alternatives.

MVP Journey Emotions:
Starts hopeful → becomes frustrated → ends disappointed or resigned

SLC Journey Emotions:
Starts curious → becomes satisfied → ends loyal and proud

Common SLC Implementation Mistakes

Even when founders embrace SLC thinking, they often make predictable mistakes:

Mistake 1: Confusing Complete with Comprehensive

Complete doesn't mean having every possible feature. It means solving the customer's job end-to-end within your chosen scope. A comprehensive social media tool might handle posting, analytics, community management, advertising, and influencer outreach. A complete social media tool might handle just posting and basic analytics, but do both so well that users don't need anything else for their specific use case.

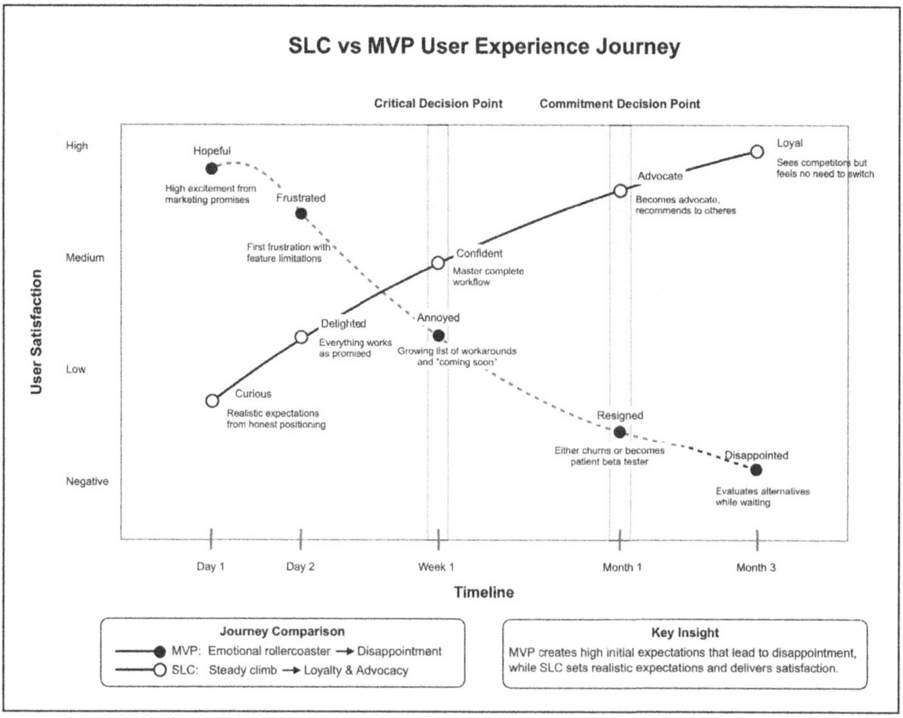

The key is choosing your scope deliberately and then being complete within that scope.

Mistake 2: Making Simple Mean Dumbed-Down

Simple doesn't mean basic. It means intuitive and clear. Some problems require sophisticated solutions, but those solutions can still feel simple to use. Professional video editing software can be simple for professionals while being powerful enough for complex projects.

Don't confuse your users' sophistication level with their desire for simplicity. Even experts prefer tools that feel effortless to use.

Mistake 3: Building Love Through Gimmicks

Love isn't about cute animations or clever copy (though those can help). Love comes from making users successful and helping them feel good about that success. Focus on meaningful delight that reinforces their achievements rather than superficial charm that distracts from their goals.

The best lovable features often feel invisible because they seamlessly support user success.

Mistake 4: Perfectionism Paralysis

SLC doesn't mean your product has to be perfect before launching. It means your product should be complete enough that users can achieve their desired outcome without frustration. You can launch an SLC product and still iterate, but you're iterating from a position of completeness rather than working toward it.

The difference is that SLC products generate satisfied customers from the very start, whereas MVP products produce patient customers who may become satisfied over time.

Measuring SLC Success

SLC products require different metrics than MVP products:

Traditional MVP Metrics:

- User activation rate
- Feature adoption
- Monthly active users
- Feedback volume

SLC Metrics:

- Job completion rate: What percentage of users successfully complete their intended job?
- Emotional satisfaction: How do users feel after using your product?
- Recommendation likelihood: Net Promoter Score and organic referral rates
- Return usage: Do users come back because they have to or because they want to?
- Competitive switching: Do users abandon alternatives after trying your product?

The goal is to measure not just usage but success and satisfaction.

The SLC Decision Framework

Not every product should be built using SLC principles. Here's when SLC makes the most sense:

Choose SLC when:

- You're entering a competitive market where users have existing alternatives
- User experience and emotion are key differentiators

- You can define a clear, valuable job that users are struggling to complete
- You have enough resources to build a complete initial solution
- Your users value quality over speed-to-market

Stick with MVP when:

- You're creating an entirely new market category
- You have extreme resource constraints
- Speed to market is your primary competitive advantage
- You're targeting early adopters who value innovation over completeness

Most established markets favor SLC approaches, while emerging markets might still benefit from MVP testing.

From SLC to Market Leadership

Companies that master SLC often achieve something that MVP products struggle with. Market leadership that's difficult to displace. When you build complete, lovable solutions, you create customers who don't want to switch even when competitors offer more features or lower prices.

This customer loyalty becomes a competitive moat that's hard to replicate. Competitors can copy your features, but they can't copy the complete experience and emotional connection you've built.

The SLC approach transforms products from feature lists into experiences worth defending. And in today's saturated markets, that might be the difference between thriving and merely surviving.

But SLC isn't the only alternative to traditional MVP thinking. In the next chapter, we'll explore the broader spectrum of modern product approaches, including Minimum Lovable Products (MLP) and Minimum Marketable Products (MMP), and how to choose the right approach for your specific situation.

The revolution in product development isn't about abandoning speed or iteration. It's about optimizing for *customer success* rather than *development convenience*. SLC is one powerful way to make that shift, but it's part of a larger movement toward building products that people don't just use, but love.

Your customers are waiting for products that respect their intelligence, solve their complete problems, and make them feel successful.

The question is: Are you ready to build one?

3

The Spectrum of Modern Product Approaches

I keep a folder on my computer labeled "Cautionary Tales" filled with pitch decks and articles about failed startups. Each one tells the same story: promising vision, strong team, adequate funding, and spectacular failure. Some tried to build everything at once and ran out of money. Others shipped too early and never recovered from poor first impressions. What strikes me most about these failures is how each company believed they had chosen the right approach. They didn't fail because they were lazy or incompetent. They failed because they picked the wrong methodology for their specific situation.

The truth is, SLC isn't the only alternative to traditional MVP thinking. Depending on your market, resources, and goals, you might need a different approach entirely. This chapter explores the full spectrum of modern product development methodologies, from Minimum Lovable Products that prioritize emotional connection with the customers to Minimum Marketable Products designed to generate revenue from day one.

Think of this as your product development toolkit. Just as a craftsperson chooses different tools for different jobs, successful founders choose different development approaches for different situations. The key is knowing which tool to use when.

Before we go further, it's worth noting that you'll come across two similar terms in product development circles: Minimum Marketable Product (MMP) and Minimum Sellable Product (MSP). Both focus on generating immediate revenue from the first release. MMP highlights the marketing point of view — creating a product that can be effectively positioned and promoted in the market. MSP emphasises the sales angle - building something that sales teams can actually sell and customers will be willing to pay for. While the terminology is slightly different, both approaches share the same core idea: prioritising features that directly contribute to revenue from day one. For clarity and consistency, this book uses MMP throughout, but when you see "MMP" referenced in later chapters, understand that it covers both MMP and MSP, the marketing and selling aspects of revenue-focused product development.

The Product Development Spectrum

Before we dive into specific methodologies, let's understand how they relate to each other. Imagine a spectrum with three primary dimensions:

Scope
How much of the customer's complete job does your product handle?
Polish
How refined and delightful is the user experience?
Speed
How quickly can you get to market?

Traditional MVP sits at one end with narrow scope, minimal polish, maximum speed. On the other hand, the SLC framework we covered in the last chapter has medium-high scope, high polish, and medium speed.

The three other modern approaches we'll explore now each target different points on this spectrum, aiming for various results depending on market conditions and business goals.

Minimum Lovable Product (MLP): When Emotion Drives Adoption

The Minimum Lovable Product concept was popularized by Brian de Haaff, CEO of Aha!, who argued that many MVP products fail not because they lack functionality, but because they fail to create emotional appeal and customer delight.

MLP asks a simple question: "What's the minimum we can build that people will genuinely love using?"

The Psychology Behind MLP

Here's something that sounds obvious but gets forgotten in the rush to ship: people are emotional creatures, even in business settings. The software you choose for your team isn't just a functional decision; it's a reflection of your taste, values, and standards. Nobody wants to be the person who recommended the ugly, clunky tool that everyone hates using.

MLP recognizes that emotional resonance often trumps feature completeness. Users will forgive missing features if they love using what's there. But they rarely forgive products that work perfectly but feel terrible to use.

Consider the early success of Medium. When it launched, it was missing basic features that competing blogging platforms had: comments, customization options, and analytics dashboards. But the writing experience was so delightful (beautiful typography, distraction-free interface, smart formatting) that writers flocked to it despite the limitations.

Medium succeeded because they understood something vital: writers don't just want to publish content. They want to feel like thoughtful, articulate authors. The emotional experience of writing on Medium made users feel more creative and professional, even when the platform couldn't do half of what WordPress could.

The MLP Design Principles

Creating lovable products isn't about adding cute animations or clever copy (though those can help). It's about understanding the emotional job your product is hired to do and designing every interaction to reinforce positive feelings.

Principle 1: Identity Reinforcement.

Great MLPs make users feel like better versions of themselves. Figma doesn't just enable design collaboration; it makes designers feel more professional and organized. Notion doesn't just organize information; it makes users feel like productivity experts with sophisticated systems.

Ask yourself: Who does your user want to be when they're using your product? How can every interaction reinforce that identity?

Principle 2: Effortless Competence.

Lovable products make difficult things feel easy, and easy things feel magical. Users should feel smart and capable, never confused or frustrated. This requires ruthless attention to the learning curve and interaction design.

Stripe revolutionized online payments not by adding more features than PayPal, but by making payment integration feel effortless for developers. Their API documentation was beautiful, their error messages were helpful, and their developer experience made programmers feel competent rather than confused.

Principle 3: Progressive Delight.

The best MLPs get more delightful the more you use them. Initial interactions should be satisfying, but long-term use should reveal hidden depths and thoughtful details that create ongoing appreciation.

Slack mastered this with features like custom emoji, slash commands, and workflow integrations. New users loved the basic messaging experience, but power users discovered layers of customization that made them feel like Slack experts.

When MLP Makes Sense

MLP works best in these situations:

Crowded Markets with Functional Parity:
When existing solutions solve the fundamental problem adequately, emotional differentiation becomes crucial. You need users to switch not because your product does more, but because it feels better.

Creative or Personal Use Cases:
Products used for creative work, personal productivity, or self-expression benefit enormously from emotional design. Users form stronger attachments to tools that make them feel creative, organized, or expressive.

Community-Driven Adoption:
In markets where word-of-mouth drives growth, lovable products spread faster than functional ones. People recommend products they love, not just products that work.

Premium Positioning:
If you plan to charge premium prices, emotional appeal justifies the cost difference. Users will pay more for products that make them feel good about their choice.

MLP Success Story: Warby Parker

Warby Parker entered the eyewear market with a simple insight: buying glasses shouldn't feel like a medical transaction. The incumbent players (Luxottica and its subsidiaries) treated eyewear as a commodity. Warby Parker treated it as a fashion and lifestyle choice.

Their MLP approach made the entire experience delightful:

Lovable Discovery: The home try-on program turned shopping into play. Customers explored styles leisurely at home, not under store pressure.

Lovable Design: Stylish, accessible glasses made wearers feel fashionable, not clinical.

Lovable Values: The company's social mission, “Buy a pair, give a pair”, made customers feel good about more than just the product.

Lovable Service: From packaging to customer support, every touchpoint reinforced the feeling that customers had made a smart, stylish choice.

Warby Parker didn't revolutionize lens technology or invent new frame materials. They made buying glasses feel delightful rather than transactional. That emotional shift created billions in value.

Minimum Marketable Product (MMP): Building to Sell from Day One

Where MLP optimizes for emotional engagement, Minimum Marketable Product optimizes for revenue generation. MMP asks: "What's the minimum we can build that customers will actually pay for, not just use?"

This distinction is crucial because paying customers behave very differently from free users. Free users will tolerate incomplete solutions, missing features, and rough edges. While paying customers expect value that justifies their investment.

The Revenue Reality Check

Here's an uncomfortable truth many founders realise too late: users who enjoy your free product won't automatically pay for your premium version. The psychology of paying creates different expectations and evaluation criteria.

When someone pays for your product, they're not just buying functionality. They're buying:

- **Reliability**: It needs to work consistently.
- **Completeness**: It needs to solve their problem end-to-end.
- **Support**: They expect help when things go wrong.
- **Professionalism**: It needs to make them look good to their colleagues/ clients.
- **ROI**: The value needs to clearly exceed the cost.

MMP thinking forces you to consider these purchasing criteria from the beginning rather than hoping to bolt them on later. MMP thinking forces you to consider these purchasing criteria from the beginning rather than hoping to bolt them on later.

This approach works across two complementary dimensions:

Marketing-Focused MMP
Creates products that can be effectively positioned and promoted in the market, with clear value propositions that resonate with target audiences.

Sales-Focused MMP
Creates products that sales teams can confidently sell and customers are willing to pay for, focusing on sales speed and revenue growth.

While marketing-focused MMP views marketability as just one factor among many, sales-focused MMP purely aims for sales velocity and revenue generation. It asks: "What's the *minimum* we can build that *maximises* immediate sales?"

The MMP Mindset

MMP thinking prioritizes features based on their impact on purchasing decisions rather than user satisfaction or technical elegance. This can feel uncomfortable for product-focused founders, but it's often necessary for business survival.

Sales-Driven Feature Prioritization
Features are ranked by how much they increase close rates, deal sizes, or sales cycle speed.

Buyer Persona Optimization
Design decisions prioritize the needs and preferences of people who sign checks, even if they're different from daily users.

Competitive Battlecards
Every feature should help win head-to-head sales battles against specific competitors.

ROI Calculators
The product should make it easy for buyers to calculate and justify the return on investment.

The MMP Development Process

Building an MMP requires a different mindset than building an MVP or MLP:

Start with Value Pricing
Instead of building features and then figuring out pricing, start with the price point that makes business sense and work backward to the value you need to deliver.

Focus on Complete Workflows
Paying customers need to accomplish complete jobs, not just try individual features. Map out the full workflow and ensure your product handles it end-to-end.

Build for Decision Makers
Free products can appeal to users who don't control budgets. Paid products need to appeal to people who sign checks, even if different people use the product daily.

Design for Professional Use
Even consumer products that people pay for need to feel substantial and reliable. The bar for "good enough" is much higher when money changes hands.

Sales-Driven Feature Prioritization
Features are ranked by how much they increase close rates, deal sizes, or sales cycle speed. Every feature should help win head-to-head sales battles against specific competitors.

When MMP Makes Sense

MMP is particularly effective in:

B2B Markets with Long Sales Cycles
When deals take months to close, customer acquisition costs are high and involve multiple stakeholders, optimizing for purchasing criteria can dramatically improve win rates.

Competitive Displacement Scenarios
When you're trying to unseat entrenched competitors, focusing on sales-winning features can help you win individual deals even if your overall product is less mature.

High-Touch Sales Models
In markets where sales teams heavily influence product adoption, building for sales effectiveness can multiply your revenue per salesperson.

Resource-Constrained Environments
When you need revenue quickly to survive or fund further development, MMP helps you generate cash flow while building toward longer-term product goals.

MMP Success Story: ConvertKit (now Kit)

Nathan Barry launched ConvertKit in an email marketing market dominated by free and inexpensive options like MailChimp. Instead of competing on price or trying to match every competitor's features, he created an MMP specifically for professional content creators who were willing to pay for better results.

ConvertKit launched with higher prices than most competitors but offered features that professional bloggers and creators needed: advanced segmentation, automation workflows, and integration capabilities. While MailChimp appealed to anyone who needed to send emails, ConvertKit focused specifically on creators who made money from their content and could justify paying for better tools.

Every feature was evaluated based on whether it would help customers make more money from their email lists, not just whether it would be cool or useful. The product and marketing consistently positioned ConvertKit as a professional tool for serious creators, not a hobby platform for casual users.

The result? ConvertKit grew from zero to millions in revenue by focusing relentlessly on building something that professional content creators would pay for, rather than something that casual email senders would use for free.

MMP Success Story: Square

Square entered the payment processing market with a deceptively simple insight: small businesses were losing sales because they couldn't accept credit cards. The existing payment infrastructure was designed for large merchants with complex needs and high transaction volumes.

Square's MMP approach focused on making credit card acceptance immediately marketable to small businesses with marketable hardware (the iconic Square card reader), marketable pricing (transparent, flat-rate pricing), marketable setup (approval in minutes rather than weeks), and marketable integration (worked with devices business owners already had).

Square didn't try to match the feature complexity of enterprise payment systems. Instead, they built exactly what small businesses needed to start accepting cards immediately and generate more revenue. The focus on marketability over feature parity allowed them to capture a massive, underserved market that traditional payment processors had ignored.

MMP Success Story: Zappos

Zappos is often miscast as an MVP story, but Nick Swinmurn's approach was pure MMP—sales-first in an existing market. He wasn't proving people would buy shoes online (they already were). He built the minimum needed to generate immediate revenue.

The sales machine prioritized:

- **Inventory-free sales:** Photographed local store shoes, bought only after orders.
- **Sales-optimized selection:** Popular styles most likely to convert.
- **Revenue-first policies:** Generous returns eliminated the #1 online shoe obstacle.

Unlike MVPs chasing insights, Zappos measured revenue from launch. Those early dollars funded their legendary customer service evolution. Success was measured in revenue, not insights. The revenue-first strategy funded their evolution into the customer service powerhouse they became recognised for.

Decision Framework: Which Approach to Choose

With multiple approaches available, how do you choose the right one for your situation? The Product Approach Decision Matrix (on the next page) is a framework based on market conditions, business goals, and resource constraints.

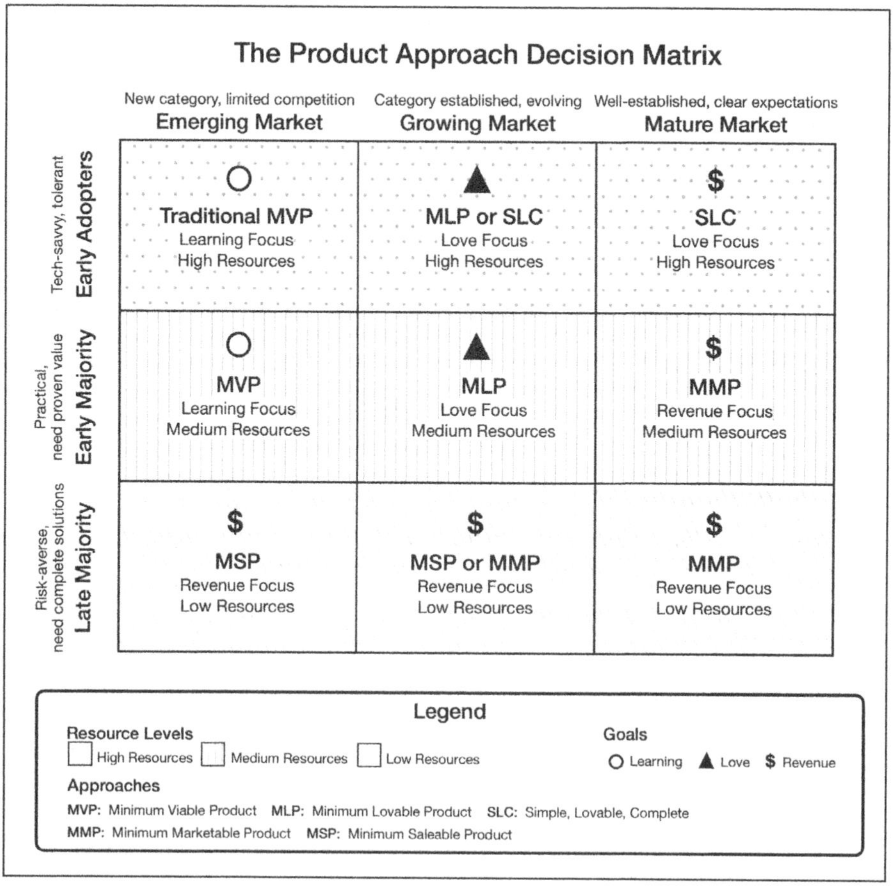

The Product Approach Decision Matrix Explanation

The Matrix is a decision tree with four key variables that determine the best approach:

Market Maturity (horizontal axis):

- **Emerging Market**: New category, limited competition, customer education needed.
- **Growing Market**: Category established but rapidly evolving, multiple viable approaches.
- **Mature Market**: Well-established category, clear customer expectations, strong incumbents.

Customer Sophistication (vertical axis):

- **Early Adopters**: Tech-savvy, willing to try new things, tolerant of limitations.

- **Early Majority**: Practical buyers, need proven value, compare alternatives carefully.
- **Late Majority**: Risk-averse, need complete solutions, follow established patterns.

Resource Level:

- **High Resources**: Can invest in completeness and polish upfront.
- **Medium Resources**: Must choose between speed and quality.
- **Low Resources**: Must optimize for the fastest path to validation/revenue.

Primary Goal (shape coding):

- **Learning**: Discovery and validation (circles).
- **Love**: Emotional engagement and retention (hearts).
- **Revenue**: Sales and cash flow (dollar signs).

The matrix provides a clear decision framework for choosing the right product approach based on your specific context and constraints.

The Key Recommendations are:

- **Traditional MVP**
  Is best for emerging markets with early adopters focused on learning.
- **MLP/SLC**
  Are optimal for growing markets targeting love and engagement.
- **MMP/MSP**
  This approach is required for mature markets and late majority customers focused on revenue.

Resource constraints will drive different approaches even within the same market/customer combination.

The Market Maturity Factor

Emerging Markets
When you're creating a new product category, traditional MVP often makes sense. Customers can't compare you to established alternatives because none exist. Your job is to validate that the category itself has merit.

Example: When Uber launched, ride-sharing didn't exist. They could test core assumptions (will people get in cars with strangers?) with a minimal product because there was no baseline for comparison.

Growing Markets
These markets offer the most flexibility. You can choose approaches based on your specific advantages and target customer segments. This is where MLP and SLC approaches often excel.

Example: When Notion entered the productivity software space, the market was growing rapidly with room for differentiated approaches. They chose an SLC-like approach, building a complete workspace rather than just another note-taking app.

Mature Markets
Established markets with strong incumbents often require MMP or MSP approaches. Customers have clear expectations and viable alternatives. You need to be obviously better or cheaper to win.

Example: When Zoom entered video conferencing, they faced entrenched competitors like Skype and WebEx. They succeeded with an MMP approach, building a complete solution that was superior for business meetings.

The Customer Sophistication Factor

Early Adopters
These customers are willing to work with incomplete products in exchange for innovation. They often prefer being first to being right. MVP and MLP approaches can work well.

Early Majority
These customers need proven value and clear differentiation. They'll compare alternatives carefully. SLC and MMP approaches tend to work better.

Late Majority
These customers are risk-averse and want complete, proven solutions. They follow rather than lead. The MMP approach is usually necessary.

Resource Constraints Reality

Your available resources often constrain your options more than market factors:

High Resources
You can choose based on optimal strategy rather than constraint. This usually means SLC or MMP approaches that invest in completeness upfront.

Medium Resources
You must choose between speed and quality. Consider MLP for emotional differentiation or focused MMP for specific market segments.

Low Resources
You're forced toward MVP for learning or MMP for quick revenue. Focus on the minimum that achieves your most critical goal.

Hybrid Approaches: Mixing and Matching

In practice, successful companies often combine elements from different approaches rather than following one methodology religiously. Here are some effective hybrid strategies:

The Progressive MLP
Start with MLP emotional design but plan a roadmap toward SLC completeness. This allows you to create emotional engagement early while building toward comprehensive solutions.

Example: Early Instagram was essentially an MLP (beautiful photo filters with social sharing) that evolved toward a more complete social media experience.

The Focused MMP
Build an MMP for a specific customer segment while planning expansion to adjacent segments. This generates revenue quickly while providing funding for broader market expansion.

Example: Salesforce started as an MMP for sales teams but expanded into a complete customer relationship platform serving multiple departments.

The Validated SLC
Use MVP techniques to validate core assumptions, then build an SLC for your proven market. This combines the learning benefits of MVP with the customer satisfaction benefits of SLC.

Example: Dropbox validated demand with a video (MVP-style testing), then built a complete file sync and sharing solution (SLC-style execution).

Common Multi-Approach Mistakes

Approach Confusion
Trying to be an MVP and an MLP simultaneously often results in products that are neither minimal nor lovable. Choose one primary approach and execute it well.

Premature Switching
Moving from MVP to MMP before achieving product-market fit often leads to products that are more polished but still don't solve real problems.

Resource Mismatch
Choosing approaches that require more resources than you have leads to half-executed strategies. Be honest about your constraints.

Market Mismatch
Using early-stage approaches (MVP) in mature markets or mature-market approaches (MMP) in emerging categories often fails.

Measuring Success Across Approaches

Different approaches require different success metrics:

MVP Metrics	SLC Metrics	MLP Metrics	MMP Metrics
• Learning velocity (insights per week) • Assumption validation rate • Iteration speed • User feedback quality	• Job completion rates • Customer satisfaction scores • Retention across user segments • Expansion revenue • Competitive win rates	• Net Promoter Score • User retention curves • Organic sharing/ referral rates • Time spent in the product • User-generated content/advocacy	• Sales conversion rates • Revenue growth rate • Average deal size • Sales cycle length • Sales velocity • Deal win percentage • Customer lifetime value • Cost of customer acquisition • Market share growth • Time to revenue

Evolution Pathways

Most successful products evolve through multiple approaches over time:

Common Evolution Path 1:

MVP → MLP → SLC

Start with learning, add emotional engagement, then build completeness.

Common Evolution Path 2:

MMP → SLC

Start with a marketable/sellable solution for a specific segment, then expand to a complete solution for a broader market.

Common Evolution Path 3:

MSP → MMP → SLC

Start with revenue focus, add marketing sophistication, then build a comprehensive solution.

The key is planning your evolution path rather than hoping it happens naturally.

Choosing Your Path Forward

As you consider which approach to take, remember that the "right" choice depends more on your specific situation than on abstract principles. Consider these questions:

1. **What's your primary constraint**: time, money, or market knowledge?
2. **What stage is your market**: emerging, growing, or mature?
3. **Who makes purchasing decisions** in your target market?
4. **What's your competitive advantage**: innovation, emotion, or execution?
5. **What does success look like** in 12 months?

Your answers should guide you toward the approach that gives you the best chance of achieving your specific goals with your available resources.

In the next chapter, we'll explore more about the emotional and psychological sides of product development, looking at how to create emotional connections that boost adoption, retention, and advocacy no matter which development approach you choose.

But first, choose your approach. Not the one that sounds most appealing or theoretically ideal, but the one that best fits your market, customers, and constraints. Because the best product development methodology is the one you can execute successfully with the resources you have available.

Your customers expect solutions that meet their needs and expectations. The question is: which approach will get you there quickest while building a sustainable business?

4

Building for Emotional Connection

I used to think emotional design was fluff. Pure decoration slapped onto functional products to make them "prettier." Then I watched a user test that changed everything. We were testing a new product on an existing website requiring user interaction to deliver the outcome. The product worked perfectly. Every feature functioned as designed, the interface was clean, and users could complete their tasks efficiently. But when the session ended, the user sighed and said, "I guess it's fine. It just feels like doing taxes."

That comment haunted me. Here was a product that solved real problems for real people, but it made them feel like they were being punished for using it. No wonder our retention rates were terrible. People don't just choose products based on functionality anymore. They choose products that make them feel the way they want to feel about their work, their relationships, and themselves.

This chapter is about understanding the psychology behind product attachment and building emotional connections that turn users into advocates. Because in a world where every product category has dozens of viable options, the ones that win aren't just the ones that work best. They're the ones that *feel* best.

The Psychology Behind Product Attachment

Here's something that might sound touchy-feely but is rooted in neuroscience: every product interaction triggers an emotional response, whether you design for it or not. The real question isn't whether your product will make users feel something. The question is whether you'll consciously influence those emotions or let them happen by chance.

The Emotional Stack of Product Experience

Think of emotional connection as happening at three levels:

Level 1: Immediate Emotional Response.
This happens in the first few seconds of interaction. Users form instant impressions based on visual design, performance, and initial functionality. At this level, emotions are primarily about competence and trust.

Questions users are unconsciously asking:

- Does this look professional?
- Will this work reliably?
- Am I going to feel stupid using this?

Level 2: Usage Emotional Experience.
This develops during active use of the product. Users form opinions about how the product makes them feel while accomplishing tasks. At this level, emotions are about efficacy and identity.

Questions users are unconsciously asking:

- Does this make me feel capable?
- Am I getting better at my work because of this?
- What does using this say about who I am?

Level 3: Reflective Emotional Association.
This happens when users think about the product when they're not using it. At this level, emotions are about values and relationships.

Questions users are unconsciously asking:

- Am I proud to use this product?
- Would I recommend this to people I respect?
- Does this align with how I see myself?

Products that create strong emotional connections operate successfully at all three levels. They look trustworthy, feel empowering during use, and create pride in reflection.

The Identity Reinforcement Principle

The most powerful emotional connections come from products that reinforce users' desired identity. People don't just want to get things done. They want to be the kind of person who gets things done in a particular way.

A designer doesn't just want to create graphics. They want to be a creative professional who produces sophisticated, beautiful work. A project manager doesn't just want to track tasks. They want to be an organized leader who keeps teams aligned and productive. A small business owner doesn't just want to process payments. They want to be a savvy entrepreneur who provides excellent customer experiences.

When products understand and reinforce these identity aspirations, they create emotional bonds that go far beyond functional utility.

The Science of User Delight

Delight isn't random. It follows predictable psychological patterns that smart product teams can deliberately trigger. Understanding these patterns allows you to create moments of genuine joy rather than just avoiding frustration.

The Delight Formula

Psychologist Dr. Noriaki Kano identified three levels of customer satisfaction that map perfectly to product emotional design:

Basic Expectations (Hygiene Factors)
These are features that users expect to work. When they work well, users feel neutral. When they don't work, users feel frustrated. These rarely create delight, but their absence creates disgust.

Examples: Login systems, data saving, basic navigation

Performance Features (Satisfiers)
These are features where more is better. The better they work, the more satisfied users feel. These can create mild positive emotions but rarely create delight.

Examples: Faster loading times, more customization options, better search results

Excitement Features (Delighters)
These are unexpected capabilities or experiences that users didn't know they wanted. When present, they create genuine delight. When absent, users don't feel disappointed because they never expected them.

Examples: Automatic data backups users never requested, smart suggestions that save time, celebratory animations for achievements

The key insight is that delight comes from exceeding expectations in unexpected ways, not from meeting expectations more efficiently.

The Surprise and Relief Pattern

Two of the most powerful emotions for creating product attachment are surprise and relief. Both require understanding user psychology and designing intentional emotional moments.

Strategic Surprise
Effective product surprises feel magical but not random. They demonstrate that the product understands the user's context and goals better than the user expected.

Examples:

- Grammarly suggesting improvements you didn't know you needed
- Spotify creating playlists that perfectly match your mood
- Calendly automatically detecting time zone differences in scheduling

Strategic Relief
Relief happens when products eliminate anxiety or effort that users expected to experience. The key is removing friction that users didn't even realize was weighing on them.

Examples:

- Automatic form filling that works correctly
- Error messages that actually help solve problems
- Sync that happens transparently across devices

Both surprise and relief work because they shift users' emotional state from neutral or negative to positive. The brain treats this emotional shift as a reward, creating positive associations with the product.

Design Thinking for Minimal Products

Traditional design thinking often assumes you have unlimited resources to create polished experiences. But what if you're building an MLP or working within SLC constraints? How do you create emotional connection when you can't build everything you want?

The Emotional Priority Framework

When resources are limited, you need to prioritize which emotional moments matter most. Here's a framework for making those decisions:

Tier 1: First Impression Emotions.
These are the feelings users have in their first 30 seconds with your product. These emotions strongly predict whether users will continue exploring or immediately leave.

Priority emotions: Trust, competence, clarity Design investment: Professional visual design, clear value proposition, intuitive navigation

Tier 2: Core Task Emotions.
These are the feelings users have while accomplishing their primary goal with your product. These emotions predict whether users will return and recommend the product.

Priority emotions: Efficacy, progress, accomplishment Design investment: Smooth workflows, clear feedback, celebration of success

Tier 3: Mastery Emotions.
These are the feelings users develop as they become proficient with your product. These emotions predict long-term retention and advocacy.

Priority emotions: Competence, pride, identity alignment Design investment: Progressive disclosure, power user features, personalization

If you can only invest in one tier, choose Tier 1. If you can invest in two, add Tier 2. Only invest in Tier 3 after the first two are working well.

The Constrained Delight Strategy

Creating delight with limited resources requires focusing on moments rather than features. Instead of trying to make everything delightful, identify the specific moments where delight will have the most impact.

The Moment Mapping Process:

1. **Map the User Journey**: Identify every significant step in your user's workflow
2. **Identify Emotional Valleys**: Find moments where users typically feel frustrated, confused, or anxious
3. **Identify Emotional Peaks**: Find moments where users typically feel accomplished, proud, or successful
4. **Choose Your Intervention Points**: Pick 2-3 moments where small design changes could create disproportionate emotional impact
5. **Design Micro-Delights**: Create small but meaningful experiences at those intervention points

For example, a simple project management tool might choose these intervention points:

- **Moment 1**: First project creation (add celebration animation)
- **Moment 2**: Task completion (add satisfying check-off experience)
- **Moment 3**: Project completion (add team sharing feature)

Three small design investments can create emotional connection throughout the user journey without requiring massive feature development.

Creating Emotional Engagement

Emotional engagement goes beyond momentary delight. It's about creating ongoing positive feelings that keep users coming back and make them feel attached to your product over time.

The Emotional Journey Architecture

Think of your product as taking users on an emotional journey from their current state to their desired state. Your job is to design that journey intentionally rather than hoping positive emotions happen accidentally.

Stage 1: Emotional Onboarding.
Users arrive with a mix of hope and skepticism. They want to believe your product will help them, but they've been disappointed before. Your emotional onboarding should quickly shift them from skeptical to confident.

Strategies:

- Quick wins that demonstrate value immediately
- Clear progress indicators that show they're moving forward
- Encouraging language that builds confidence
- Removal of unnecessary decisions that create anxiety

Stage 2: Emotional Momentum.
Once users see initial value, your job is to maintain positive emotional momentum. Users should feel increasingly capable and successful as they use your product more.

Strategies:

- Skill progression that makes users feel like they're improving
- Social proof that reinforces their good judgment in choosing your product
- Personalization that makes the product feel increasingly tailored to them
- Achievement systems that celebrate their progress

Stage 3: Emotional Mastery.
Advanced users should feel like experts who have made a smart choice in committing to your product. They should feel proud of their proficiency and eager to share their success with others.

Strategies:

- Advanced features that make users feel powerful
- Community recognition of their expertise
- Opportunities to mentor other users
- Integration options that make your product central to their workflow

The Psychology of Product Habit Formation

The strongest emotional connections come from products that become integrated into users' daily routines. When using your product becomes habitual, the emotional connection becomes automatic and unconscious.

The Habit Loop for Product Design:

Cue: Environmental trigger that initiates product use.
Routine: The behavior of using your product.
Reward: The benefit or positive feeling from using your product.
Investment: The user action that increases future engagement

For emotional connection, the reward phase is critical. Users need to feel genuinely better after using your product than they did before. This positive emotional change is what transforms occasional usage into habitual engagement.

Designing Effective Reward Cycles:

Variable Rewards:
Like slot machines, products that provide unpredictable positive outcomes create stronger habits than products with consistent rewards.

Example: LinkedIn notifications that sometimes contain genuinely valuable opportunities mixed with routine updates.

Social Rewards:
Humans are deeply motivated by social recognition and belonging. Products that help users feel connected to others or recognized by peers create strong emotional bonds.

Example: GitHub's contribution graphs that make coding feel like a social, visible activity.

Achievement Rewards:
People are motivated by feelings of progress and mastery. Products that help users feel like they're improving create positive emotional associations.

Example: Duolingo's streak counters that make language learning feel like measurable progress.

Case Study: Slack's Lovable Simplicity

Slack revolutionized workplace communication not through technical innovation, but through emotional design that made team communication feel collaborative rather than transactional.

The Emotional Problem Slack Solved

Before Slack, workplace communication felt fragmented and frustrating. Email chains created confusion, instant messaging felt unprofessional, and important conversations got lost in the noise. Teams felt disconnected even when they were constantly communicating.

Slack understood that the emotional aspect of the job was bigger than just sending messages. Teams wanted to feel aligned, productive, and connected. They wanted communication tools that enhanced their collaborative relationships rather than creating friction.

Slack's Emotional Design Decisions

Lovable Personality
Slack's interface had personality without being unprofessional. Friendly error messages, playful loading states, and customizable emoji made work communication feel more human without sacrificing functionality.

Emotional Clarity
Channels created clear contexts for different types of conversations. Instead of drowning in undifferentiated message streams, users could organize their communication in ways that felt logical and manageable.

Social Connection
Features like status updates, custom emoji, and reaction buttons made digital communication feel more expressive and personal. Teams could develop their own communication culture within Slack.

Progress Visibility
Message threading and search made it easy to follow conversation progress and find important information. Users felt in control of their communication rather than overwhelmed by it.

The Results of Emotional Focus

Slack's emotional approach created business results that purely functional approaches couldn't match:

Viral Adoption:
Teams adopted Slack organically because using it felt better than alternatives. The emotional experience drove word-of-mouth growth.

High Retention:
Teams that started using Slack rarely switched to alternatives because the emotional connection made switching feel like a downgrade.

Premium Pricing:
Organizations paid premium prices because Slack made teams feel more productive and connected. The emotional value justified the cost.

Category Disruption:
Slack didn't compete with existing communication tools on features. They redefined what workplace communication could feel like.

The User Emotion Stack Diagram

The diagram (below) shows how each layer builds on the previous one, with failure at any layer preventing progression to higher emotional states. Most products focus only on the bottom two layers, while the most beloved products deliberately design experiences for all five layers.

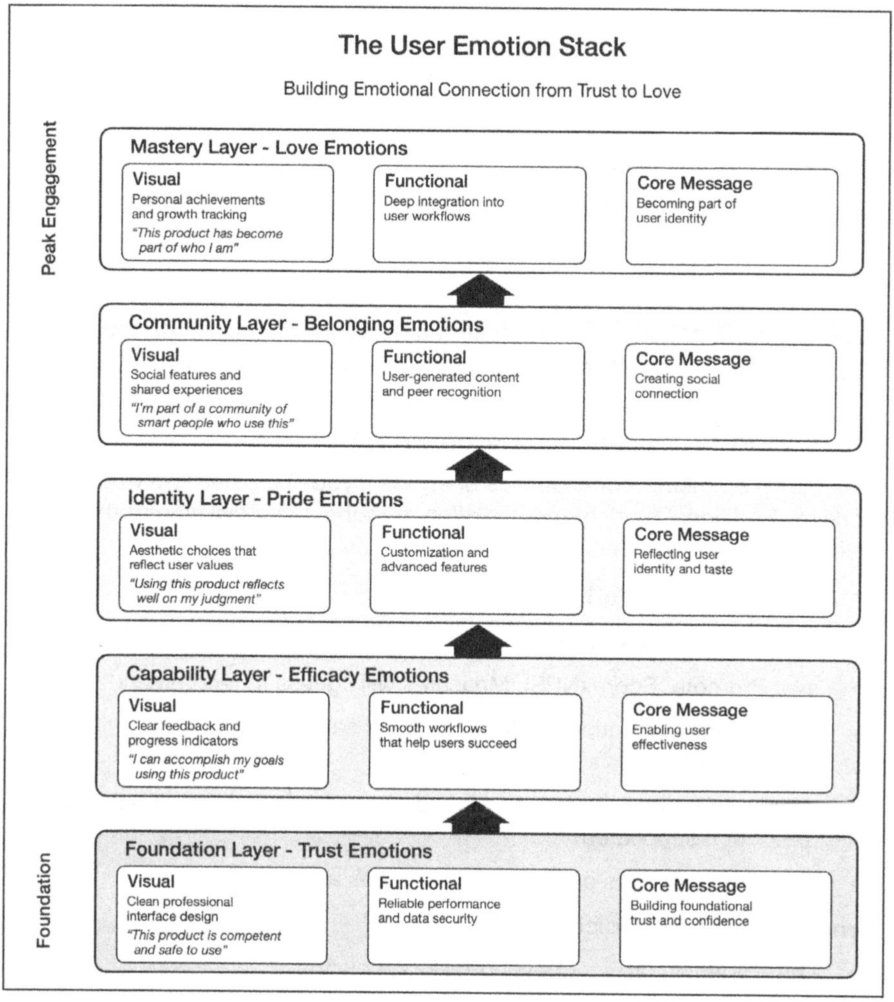

Stack Structure (Bottom to Top):

1. **Foundation Layer - Trust Emotions**:
   Building basic confidence through professional design and reliable performance

2. **Capability Layer - Efficacy Emotions**:
   Enabling user success through clear feedback and smooth workflows
3. **Identity Layer - Pride Emotions**:
   Reflecting user values through aesthetics and customization
4. **Community Layer - Belonging Emotions**:
   Creating social connection through shared experiences
5. **Mastery Layer - Love Emotions**:
   Becoming part of user identity through deep integration

Each layer shows three key aspects:

- **Visual**: The design elements that communicate the emotion
- **Functional**: The features that deliver the emotional experience
- **Core Message**: The user's internal narrative at that level

The arrows between layers show the progression from foundational trust to peak emotional engagement, with each layer building on the previous one to create a deeper user connection.

Measuring Emotional Engagement

Creating an emotional connection is pointless if you can't measure whether it's working. Traditional product metrics often miss the emotional signals that predict long-term success.

Emotional Engagement Metrics

Direct Emotion Measurement:

- Net Promoter Score (NPS): Measures willingness to recommend
- Customer Satisfaction (CSAT): Measures satisfaction with specific interactions
- Product-Market Fit surveys: "How disappointed would you be if this product disappeared?"
- Sentiment analysis of support interactions and user feedback

Behavioral Emotion Indicators:

- Time spent in product beyond task completion
- Frequency of voluntary return visits
- User-generated content creation (reviews, social posts, tutorials)
- Feature adoption rates for non-essential features
- Response rates to optional surveys and feedback requests

Social Emotion Signals:

- Organic mentions on social media
- Unprompted reviews and testimonials
- Referral rates and viral coefficients
- Community participation and peer-to-peer support
- Employee advocacy (for B2B products)

Long-term Attachment Indicators:

- Retention curves over 6+ months
- Expansion revenue from existing customers
- Resistance to competitive offers
- Recovery rates after service issues
- Participation in product beta programs

The Emotional Health Dashboard

Create a dashboard that tracks emotional engagement alongside traditional product metrics:

Weekly Emotional Pulse:

- NPS trend over time
- Support ticket sentiment analysis
- Social media mention sentiment
- User-generated content volume

Monthly Emotional Deep Dive:

- Cohort retention analysis
- Feature adoption patterns
- Community engagement metrics
- Competitive win/loss analysis

Quarterly Emotional Audit:

- Full customer satisfaction survey
- User interview themes analysis
- Churn reason categorization
- Product-market fit assessment

This emotional dashboard helps you understand not just whether your product is being used, but whether it's being loved.

Building Love into Product DNA

Emotional connection can't be bolted on after the fact. It needs to be built into your product development process from the beginning.

The Emotional Product Requirements

For every feature you build, define not just functional requirements but emotional requirements:

Functional Requirement: "Users can create and share project reports"

Emotional Requirement: "Users feel confident sharing reports that make them look competent and organized to their colleagues"

This emotional requirement changes design decisions. It's not enough for the feature to work; it needs to make users feel good about themselves when they use it.

The Emotional Design Review Process

Add emotional criteria to your design review process:

Pre-Development Review:

- What emotion do we want users to feel during this interaction?
- What could go wrong emotionally, and how do we prevent it?
- How does this feature reinforce positive user identity?

Post-Development Review:

- Does the implementation create the intended emotional experience?
- What unintended emotions might this create?
- How can we measure whether the emotional goals are achieved?

Post-Launch Review:

- Are users experiencing the emotions we designed for?
- What emotional feedback are we receiving?
- How should we adjust based on emotional data?

The Emotional Competitive Advantage

Products that create strong emotional connections develop sustainable competitive advantages that are difficult to replicate:

Emotional Switching Costs:
Users resist switching from products they love, even when alternatives offer better features or pricing.

Emotional Price Premiums:
Users pay more for products that make them feel good about their choice.

Emotional Marketing Efficiency:
Loved products benefit from organic word-of-mouth that reduces customer acquisition costs.

Emotional Innovation Direction:
Passionate users provide better feedback and product ideas because they're emotionally invested in success.

Emotional Talent Attraction:
Companies known for building beloved products attract better employees who want to work on products people love.

These advantages compound over time, creating defensible market positions that competitors struggle to attack directly.

Your Emotional Design Action Plan

Ready to build emotional connection into your product? Here's where to start:

Week 1: Emotional Audit

- Survey existing users about how your product makes them feel
- Analyze support tickets and user feedback for emotional themes
- Identify the top 3 emotions you want users to experience
- Map your current user journey and identify emotional low points

Week 2: Emotional Strategy

- Define the identity your users want to embody
- Identify 3 specific moments where you can create delight
- Establish emotional requirements for your next 3 features
- Create an emotional measurement plan

Week 3: Emotional Implementation

- Redesign your most critical user interaction with emotional goals in mind
- Add delight moments to 2 existing features
- Update your product copy to reinforce positive user identity
- Implement emotional tracking in your analytics

Week 4: Emotional Optimization

- Test emotional changes with real users
- Measure baseline emotional engagement metrics
- Plan emotional improvements for the next quarter
- Train your team on emotional design principles

Building emotional connection takes time, but the investment pays dividends in customer loyalty, word-of-mouth growth, and sustainable competitive advantage.

In the next chapter, we'll explore how to test and validate these emotional design decisions using both quantitative metrics and qualitative feedback. Because loving customers is great, but knowing how to create more loving customers is even better.

Your users are already forming emotional connections with your product, whether you design for them or not. The question is whether those emotions will help your business grow or hold it back. Choose wisely, and design with intention.

5
Testing and Validation Strategies

I once watched a startup founder spend six months driving traffic to their MVP while completely ignoring the fact that more than 80% of users abandoned the product after the first screen. When I asked why they weren't testing the fundamental user experience, they proudly showed me their traffic dashboard with different color variations for traffic sources. "We're very data-driven," they declared, apparently missing the irony that their data was telling them exactly the wrong story.

This chapter is about testing the things that matter. Whether you're building an MVP, SLC, MLP, or MMP, you need validation strategies that match your approach and measure what predicts success. Because testing the wrong things perfectly is still perfectly wrong.

The Validation Paradigm Shift

Traditional MVP testing focuses on one question: "Are we building the right thing?" But modern product approaches require more nuanced questions:

- **MVP**: Are we learning the right things?
- **SLC**: Are we delivering complete value?
- **MLP**: Are we creating emotional engagement?
- **MMP**: Are we building something people will buy?

Each approach requires different testing methodologies because they optimize for different outcomes. Using MVP testing strategies for an MLP is like using a ruler to measure temperature; you're using the wrong tool for the job.

The Testing Evolution Framework

Think of testing as evolving through four generations:

Generation 1: Feature Testing
"Does this feature work as designed?"
Focus: Functionality and bug identification
Tools: QA testing, unit tests, integration tests

Generation 2: Usability Testing
"Can users figure out how to use this feature?"
Focus: User interface and experience design
Tools: User testing, heuristic evaluation, task completion rates

Generation 3: Value Testing
"Does this feature create value for users?"
Focus: Outcomes and user success
Tools: Behavioral analytics, outcome measurement, cohort analysis

Generation 4: Emotional Testing
"Does this feature make users feel the way we want them to feel?"
Focus: Emotional response and attachment
Tools: Sentiment analysis, engagement depth, advocacy measurement

Most teams get stuck in Generation 2, endlessly refining usability while ignoring whether they're building something people actually want or love.

MVP Testing: Rapid Learning Through Behavioral Data

MVP testing should prioritize learning velocity over comprehensive validation. You're not trying to prove your product is perfect; you're trying to discover what's true about your market as quickly as possible.

The MVP Testing Mindset

Assumption-Driven Testing:
Start with a list of your riskiest assumptions and design tests specifically to validate or invalidate them.

Behavioral Over Stated:
What people do matters more than what they say they'll do. Focus on measuring actual behavior.

Directional Over Precise:
You need directional insights (is this working or not?) rather than precise measurements (exactly how well is this working?).

Speed Over Perfection:
Fast, imperfect insights beat slow, perfect ones when you're in learning mode.

Essential MVP Testing Methods

The Fake Door Test

Create landing pages or buttons for features that don't exist yet. Measure click-through rates to gauge demand before building.

Example Setup:

Add "Advanced Analytics" button to your basic dashboard

Track clicks and user reactions when they discover it's "coming soon"

If >15% of users click and >50% of those leave their email for updates, consider building it

The Concierge Test

Manually deliver the service you plan to automate. This tests whether the core value proposition works before you invest in technology.

Example Implementation:

Instead of building automated report generation, manually create reports for 10 users

Measure: Do they use the reports? How often? What questions do they ask?

Only build automation after proving manual version creates value

The Wizard of Oz Test

Build a simple interface that appears to work automatically but is actually powered by humans behind the scenes.

Example Framework:

Create a "smart recommendation engine" that's actually powered by manual curation

Test whether users value the recommendations before building AI

Measure engagement with recommended content

The Cohort Learning Test

Instead of testing individual features, test different value propositions with different user cohorts.

Example Structure:

Cohort A gets basic project management features

Cohort B gets collaboration-focused features

Cohort C gets reporting and analytics features

Measure which cohort has highest engagement and retention

MVP Testing Metrics That Matter

Leading Indicators:

- Time to first value (how quickly do users get their first success?)
- Activation rate (what percentage reach a meaningful outcome?)
- Assumption validation rate (how many core assumptions are proving true?)

Lagging Indicators:

- Weekly retention (do users come back?)
- Organic growth rate (do users bring others?)
- Problem-solution fit score (how disappointed would users be if this disappeared?)

Learning Velocity Metrics:

- Experiments launched per week
- Time from hypothesis to results
- Percentage of assumptions tested before building features

SLC Testing: Measuring Complete Value Delivery

SLC testing focuses on whether you're delivering end-to-end value within your chosen scope. You're not just testing individual features; you're testing whether the complete workflow feels effortless and valuable.

The SLC Testing Philosophy

Workflow Completeness:
Test entire user journeys, not individual features. Users care about accomplishing complete jobs, not using isolated capabilities.

Emotional Journey Mapping:
Track how users feel at each step of their journey. SLC products should create increasingly positive emotions as users progress.

Simplicity Under Pressure:
Test how your product performs when users are stressed, distracted, or in a hurry. Simple products should remain usable even under pressure.

Expert User Validation:
Test with both novice and expert users. SLC products should feel simple to beginners but powerful to experts.

SLC Testing Methodologies

The Complete Job Test

Give users realistic scenarios that require them to accomplish complete objectives using your product.

Test Structure:

- Scenario: "You need to plan and execute a successful product launch"
- Success Metrics: Can they complete the entire job using only your product?
- Emotional Metrics: How do they feel about the experience?
- Simplicity Metrics: How many times do they feel confused or stuck?

The Competitive Workflow Test

Have users accomplish the same task using your product and a competitor's product. Measure both completion rates and emotional responses.

Comparison Framework:

- Task completion time
- Error rates and recovery time
- Emotional state before and after
- Preference and reasoning

The Integration Stress Test

Test how well your product works within users' existing workflows and tool ecosystems.

Test Scenarios:

- How easily does data flow in and out of your product?
- Does your product create workflow gaps that require other tools?
- How does your product behavior integrate with users' mental models?

The Onboarding-to-Mastery Test

Track user progression from first use to power user status. SLC products should have smooth progression curves.

Progression Metrics:

- Time to basic competence
- Feature adoption curve over time
- User confidence ratings at different skill levels
- Abandonment points in the learning journey

SLC Success Indicators

Workflow Metrics:

- End-to-end task completion rates
- Time to complete complex workflows
- Number of external tools required to complete jobs
- User satisfaction with complete processes

Simplicity Metrics:

- Learning curve steepness
- Help documentation usage rates
- Support ticket volume per user
- User error recovery rates

Completeness Metrics:

- Feature request patterns (are users asking for core workflow gaps?)
- Workflow abandonment points
- External tool usage alongside your product
- User success story complexity

MLP Testing: Measuring Delight and Emotional Resonance

MLP testing goes beyond functionality to measure whether users actually love using your product. You're testing emotional responses, advocacy behavior, and long-term attachment.

The MLP Testing Approach

Emotion-First Metrics:
Traditional metrics like conversion rates and retention are important, but MLP testing prioritizes emotional engagement metrics that predict long-term success.

Qualitative Over Quantitative:
While behavioral data is important, MLP testing relies heavily on qualitative feedback about feelings, preferences, and emotional responses.

Long-term Relationship Indicators:
MLP testing focuses on metrics that predict whether users will develop lasting relationships with your product.

Community and Advocacy Signals:
Lovable products create communities and advocates. Testing should measure these organic behaviors.

MLP Testing Methods

The Emotional Journey Interview

Conduct in-depth interviews that focus on feelings rather than just functionality.

Interview Structure:

- How did you feel when you first saw the product?
- What emotions do you experience while using it?
- How do you feel after completing tasks?
- What would you tell friends about this product?
- What emotions would you feel if this product disappeared?

The Social Proof Test

Measure organic sharing and recommendation behavior without prompting or incentives.

Tracking Metrics:

- Unprompted social media mentions
- Organic referral rates
- User-generated content volume
- Community participation rates
- Net Promoter Score trends

The Competitive Love Test

Compare emotional responses to your product versus alternatives.

Test Framework:

- Users try your product and a competitor's for the same tasks
- Measure functional completion rates (hygiene factor)
- Measure emotional preferences and reasoning
- Track which product they choose for ongoing use

The Investment Behavior Test

Measure how much users are willing to invest in your product beyond basic usage.

Investment Indicators:

- Time spent on non-essential features
- Customization and personalization effort
- Community participation and content creation

- Willingness to learn advanced features
- Response to optional engagement opportunities

MLP Emotional Metrics

Immediate Emotional Response:

- First impression ratings
- Time spent exploring beyond core tasks
- Feature discovery rates
- Help-seeking behavior patterns

Ongoing Emotional Engagement:

- Session duration beyond task completion
- Return visit frequency for non-urgent tasks
- Feature adoption rates for "delight" features
- Customization and personalization usage

Long-term Emotional Attachment:

- Net Promoter Score progression over time
- Community participation rates
- User-generated content creation
- Resistance to competitive alternatives

MMP Testing: Validating Purchase Intent and Market Readiness

MMP testing focuses on whether you've built something that people will actually buy, not just use. You're testing purchasing psychology, buyer decision-making processes, and revenue potential.

The MMP Testing Priorities

Purchase Intent Over Usage Intent: Free usage doesn't predict paid usage. MMP testing measures willingness to pay, not just willingness to try.

Buyer Persona Validation: Understanding who makes purchasing decisions and what drives those decisions is crucial for MMP success.

Value Perception Testing: Testing whether your target market perceives enough value to justify your pricing and switching costs.

Sales Process Optimization: Testing and refining every step of the path from awareness to purchase.

MMP Testing Strategies

The Price Sensitivity Test

Use various price points to understand demand elasticity and optimal pricing.

Testing Framework:

- Van Westendorp Price Sensitivity Meter surveys
- A/B testing different pricing pages
- Willingness-to-pay conjoint analysis
- Competitive pricing comparison studies

The Purchase Simulation Test

Create realistic purchase scenarios without actually charging users.

Simulation Structure:

- Complete sales process including payment information entry
- Stop just before final payment processing
- Measure drop-off rates at each stage
- Interview users about purchase decision factors

The Buyer Journey Mapping Test

Understand how purchase decisions actually happen in your target market.

Research Process:

- Interview recent buyers of similar products
- Map their complete decision-making process
- Identify key decision criteria and influencers
- Test how well your product addresses each criterion

The Competitive Win/Loss Test

Test head-to-head against alternatives in realistic purchase scenarios.

Test Structure:

- Present your product alongside competitors
- Let users evaluate options for realistic use cases
- Track selection rates and reasoning
- Identify specific factors that drive purchase decisions

MMP Revenue Indicators

Purchase Intent Metrics:

- Conversion rates from trial to paid
- Price sensitivity curves
- Purchase decision timeframes
- Buyer persona accuracy

Market Readiness Metrics:

- Sales qualified lead generation rates
- Demo-to-trial conversion rates
- Trial-to-purchase conversion rates
- Average deal size progression

Competitive Positioning Metrics:

- Win rates against specific competitors
- Deal cycle length compared to alternatives
- Price premium sustainability
- Customer acquisition cost efficiency

The Testing Strategy Comparison Framework

The Framework's quadrants (next page) represent four different testing approaches:

1. **MVP Territory**:
   Quick functional validation using A/B tests, usage analytics, and task completion metrics over days to weeks
2. **MLP Territory**:
   Emotional engagement testing using NPS, sentiment analysis, and community metrics over months to quarters
3. **MMP Territory**:
   Purchase validation using conversion testing, price sensitivity, and buyer interviews over weeks to months
4. **SLC Territory**:
   Complete value delivery testing using workflow testing, retention cohorts, and job completion over months to years

Each quadrant includes the specific focus, methods, timeline, and key question that defines that testing approach. The axis arrows clearly show the dimensional

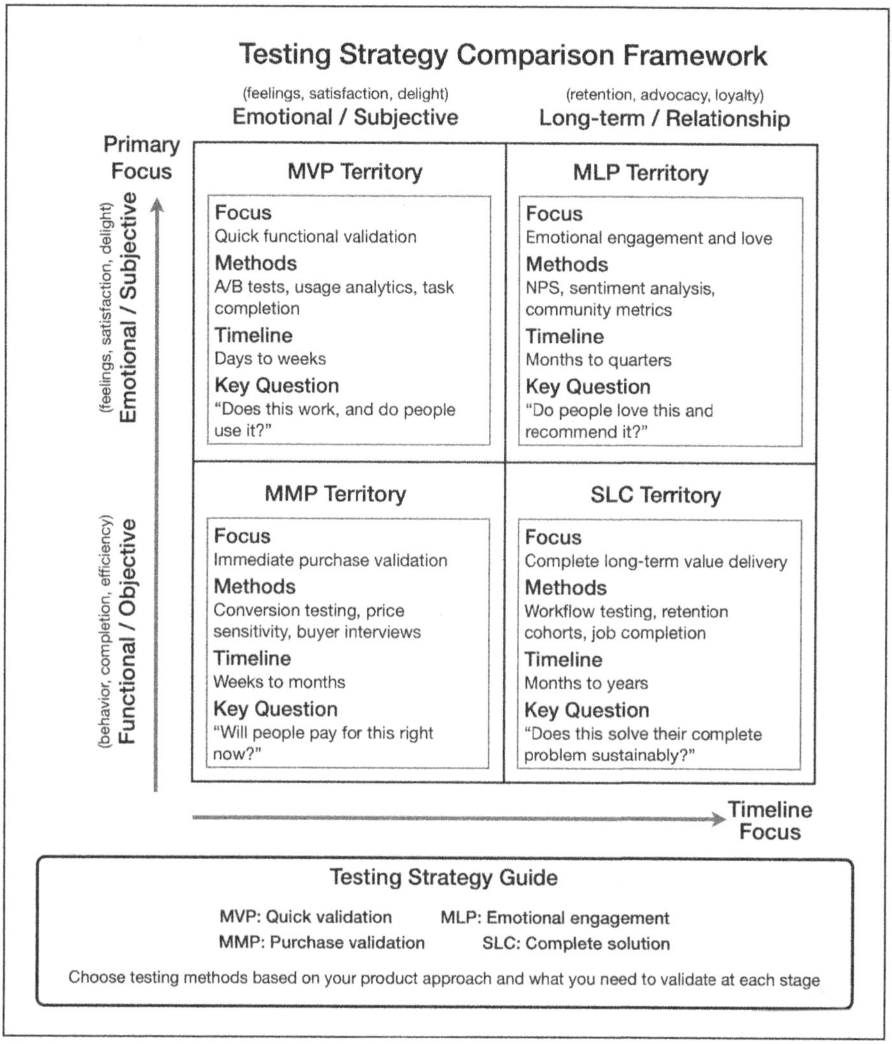

relationships between immediate vs. long-term focus and emotional vs. functional testing priorities.

Overlapping Areas: Most successful products eventually test in multiple quadrants, but start with the quadrant that best aligns with their primary approach.

Advanced Testing Techniques

Once you master basic testing for your chosen approach, these advanced techniques can provide deeper insights:

Multi-Method Triangulation

Don't rely on a single testing method. Use multiple approaches to validate the same hypothesis:

Hypothesis: "Users find our onboarding process intuitive"

Method 1: Behavioral analytics (completion rates, drop-off points)
Method 2: User testing sessions (think-aloud protocols)
Method 3: Post-onboarding surveys (perceived difficulty ratings)
Method 4: Support ticket analysis (common confusion points)

Convergent results across methods increase confidence. Divergent results reveal hidden complexity.

Longitudinal Cohort Analysis

Track the same users over extended periods to understand how relationships with your product evolve:

Week 1: Initial reaction and basic usage patterns
Month 1: Feature adoption and workflow integration
Quarter 1: Retention, satisfaction, and advocacy development
Year 1: Long-term value realization and competitive comparison

This longitudinal view reveals patterns that cross-sectional testing misses.

Competitive Intelligence Testing

Understand not just how users respond to your product, but how they compare you to alternatives:

Side-by-Side Testing: Users evaluate your product alongside competitors
Switching Studies: Understanding why users switch between products
Feature Gap Analysis: Identifying capabilities that drive competitive advantage
Brand Perception Testing: How your product positioning affects evaluation

Predictive Analytics Testing

Use early signals to predict long-term outcomes:

Leading Indicators: Early behaviors that predict retention
Cohort Progression Models: How initial usage patterns predict long-term value
Churn Prediction: Identifying users at risk before they leave
Expansion Revenue Signals: Behaviors that predict increased spending

Common Testing Mistakes and How to Avoid Them

Mistake 1: Testing Too Early

The Problem: Testing incomplete features or experiences that don't represent the intended final state.

The Fix: Ensure your test scenarios represent complete, realistic use cases. Better to test fewer scenarios well than many scenarios poorly.

Mistake 2: Testing the Wrong People

The Problem: Using convenient test subjects who don't represent your real target market.

The Fix: Invest in recruiting test participants who match your ideal customer profile. Wrong participants lead to wrong conclusions.

Mistake 3: Measuring Vanity Metrics

The Problem: Tracking metrics that feel good but don't predict business success.

The Fix: Connect every test metric to business outcomes. Ask "If this metric improves, what business result should follow?"

Mistake 4: Over-Testing Micro-Decisions

The Problem: Spending too much time testing small details while ignoring major user experience issues.

The Fix: Use the 80/20 rule. Focus testing effort on the 20% of decisions that drive 80% of user experience.

Mistake 5: Confusing Statistical Significance with Practical Significance

The Problem: Acting on test results that are statistically reliable but practically meaningless.

The Fix: Define minimum effect sizes that would change your product decisions before running tests.

Building a Testing Culture

Effective testing isn't just about methods; it's about building organizational habits that prioritize learning and user understanding.

The Testing Mindset

Hypothesis-Driven Development:
Every feature should start with a clear hypothesis about user behavior and value creation.

Embrace Negative Results:
Failed tests that invalidate assumptions are as valuable as successful tests that confirm them.

Iterative Improvement:
View testing as an ongoing practice, not a one-time validation exercise.

Cross-Functional Collaboration:
Include design, engineering, marketing, and customer success in testing planning and analysis.

Testing Process Integration

Pre-Development Testing: Validate concepts before building
Development Testing: Validate implementation during building
Post-Launch Testing: Validate outcomes after shipping
Continuous Testing: Ongoing monitoring and optimization

Your Testing Implementation Plan

Week 1: Testing Audit

- Identify which approach (MVP/SLC/MLP/MMP) best describes your product
- Map your current testing methods against the framework
- Identify gaps between your approach and your testing strategy
- Choose 3 new testing methods to implement

Week 2: Testing Setup

- Implement analytics for your chosen approach's key metrics
- Create testing templates for your most important scenarios
- Recruit a pool of test participants who match your target market
- Establish baseline measurements for key metrics

Week 3: Testing Execution

- Run your first round of approach-specific tests
- Gather both quantitative data and qualitative feedback
- Document insights and surprises
- Plan iteration based on results

Week 4: Testing Optimization

- Analyze patterns across multiple test results
- Refine your testing process based on what you've learned
- Plan ongoing testing schedule and resource allocation

- Share insights with your team and stakeholders

Remember: The best testing strategy is the one that matches your product approach and gets implemented. Perfect testing that doesn't happen is infinitely less valuable than good-enough testing that provides regular insights.

In the next chapter, we'll explore how different approaches affect resource allocation and development planning because choosing the right testing strategy is pointless if you don't have the resources to execute it effectively.

Your users are constantly testing your product, whether you're measuring their responses or not. The question is whether you'll learn from their tests before or after they decide to stop using your product.

6
Resource Allocation and Development Planning

Here's a pattern that shows up repeatedly in my work. I'll review a product launch post-mortem, and the team will describe their eight-month development cycle that burned through their seed funding. When I ask what they built, they'll outline a feature-rich platform with custom integrations, advanced analytics, and multiple client applications. Then someone will add, "But we called it an MVP," as if the label somehow justified the reality of the scope creep and resource allocation. The disconnect between what teams build and what they call it reveals a fundamental misunderstanding of product development approaches.

This chapter is about the uncomfortable truth that your development approach determines your resource requirements, not the other way around. You can't build an SLC product on an MVP budget, and you can't build an MLP with a purely engineering team. Understanding these resource implications upfront will save you from painful discoveries later.

The Resource Reality Check

Different product approaches don't just require different features; they require fundamentally different resource allocation strategies. The mistake many founders make is choosing an approach based on what sounds good, then trying to force their resources to fit rather than choosing an approach that matches their actual constraints.

The Four Resource Dimensions

Time:
How long can you afford to spend before you need market feedback or revenue?

Money:
What's your total budget for reaching your next milestone?

Team:
What skills and capacity do you actually have available?

Opportunity Cost:
What else could you be doing with these resources?

Each product approach makes different trade-offs across these dimensions. Understanding these trade-offs helps you choose an approach that's actually executable with your resources.

The Hidden Cost Multipliers

What makes resource planning tricky is that certain decisions create hidden multipliers that dramatically increase resource requirements:

Polish Multiplier:
Making something feel "lovable" rather than just "functional" typically adds 50-100% to development time.

Completeness Multiplier:
Building end-to-end workflows rather than individual features can add 100-200% to the scope.

Integration Multiplier:
Making systems work together seamlessly often takes longer than building them separately.

Quality Multiplier:
Building for paying customers requires higher reliability and support standards than building for free users.

These multipliers aren't inherently bad, but they need to be planned for rather than discovered during development.

MVP Resource Allocation: Speed Over Everything

MVP resource allocation optimizes for learning velocity and speed to market. In the MVP world, everything else is secondary to getting something testable in front of users quickly.

The MVP Resource Philosophy

Time Constraints Drive Scope:
You start with a fixed timeline (usually 4-8 weeks) and cut scope to fit rather than extending time to fit scope.

Minimal Team Configuration:
Small, cross-functional teams that can make decisions quickly without extensive coordination.

Technical Debt Acceptance:
MVP teams deliberately take on technical debt to achieve speed, planning to refactor later based on learning.

Quality Threshold:
"Good enough to test" rather than "good enough to sell" or "good enough to love."

MVP Team Structure

The Classic MVP Team (3-4 people):

- 1 Product/Business person (requirements, testing, user research)
- 1-2 Engineers (building, basic design, technical decisions)
- 1 Designer/UX person (user experience, visual design, usability)

This small team size enables rapid decision-making and reduces coordination overhead. Everyone needs to be comfortable wearing multiple hats and making imperfect decisions quickly.

MVP Budget Allocation

70% Engineering: Building the core functionality

20% Design/UX: Making it usable enough to test

10% Other: Infrastructure, tools, validation activities

MVP budgets prioritize building over everything else. Marketing, sales, and customer success are minimal or non-existent.

MVP Timeline Planning

Week 1-2: Requirements definition and technical architecture

Week 3-6: Core feature development

Week 7: Basic testing and bug fixes

Week 8: Launch and initial user feedback collection

MVP timelines are aggressive and require ruthless scope management. Features that don't directly test core assumptions get cut.

When MVP Resource Allocation Works

Clear Learning Objectives:
You have specific hypotheses to test and clear criteria for success/failure.

Low Integration Requirements:
You're building something relatively standalone that doesn't require complex system integration.

Forgiving User Base:

Your test users understand they're using an experimental product and won't judge you harshly for limitations.

Strong Technical Team:
Your engineers can build quickly without extensive oversight or process.

When MVP Resource Allocation Fails

Quality Expectations:
If your users expect production-quality software, MVP resource allocation will disappoint them.

Complex Problem Domain:
Some problems require significant upfront investment to understand and solve properly.

Compliance Requirements:
Regulated industries often require quality standards that conflict with MVP speed priorities.

Team Learning Curve:
If your team needs to learn new technologies or domains, MVP timelines become unrealistic.

SLC Resource Allocation: Balanced Investment for Complete Solutions

SLC resource allocation balances speed with completeness, requiring more upfront investment but targeting sustainable, long-term success.

The SLC Resource Philosophy

Quality-Speed Balance:
SLC accepts longer development times in exchange for higher quality initial releases.

Complete Workflow Investment:
Resources are allocated to ensure users can accomplish complete jobs, not just use individual features.

Design-Engineering Parity:
Equal investment in user experience design and technical implementation.

Sustainable Technical Foundation:
Building for long-term maintainability rather than short-term speed.

SLC Team Structure

The Balanced SLC Team (5-7 people):

- 1 Product Manager (requirements, strategy, user research)

- 2-3 Engineers (feature development, architecture, quality)
- 1-2 Designers (UX research, interface design, user testing)
- 1 QA/Testing person (quality assurance, user acceptance testing)

SLC teams are larger and more specialized than MVP teams, enabling higher quality output but requiring more coordination.

SLC Budget Allocation

50% Engineering: Technical implementation and architecture

30% Design/UX: User experience, interface design, user research

10% Quality Assurance: Testing, validation, bug fixing

10% Other: Infrastructure, tools, project management

SLC budgets invest heavily in design and quality to ensure the complete experience feels polished.

SLC Timeline Planning

Week 1-3: User research, requirements definition, technical architecture

Week 4-10: Iterative development with continuous user testing

Week 11-12: Integration testing and polish

Week 13-14: Launch preparation and final validation

SLC timelines are longer but more predictable than MVP timelines. The additional upfront planning reduces surprises during development.

SLC Resource Success Factors

Clear Job Definition:
You understand the complete job users are trying to accomplish and can scope a complete solution.

Design-Engineering Collaboration:
Your team can work together effectively across disciplines.

User Access:
You have ongoing access to target users for research and testing throughout development.

Quality Standards:
Your market expects polished experiences and will judge incomplete solutions harshly.

MLP Resource Allocation: Design-Heavy Investment for Emotional Impact

MLP resource allocation prioritizes user experience design and emotional engagement, often requiring design skills and sensibilities that traditional teams lack.

The MLP Resource Philosophy

Design-Led Development:
Design drives technical requirements rather than being constrained by them.

Emotional Experience Investment:
Significant resources dedicated to creating delight and emotional connection.

Iterative Polish:
Multiple rounds of refinement to achieve the right emotional impact.

User Research Intensity:
Heavy investment in understanding user emotions and motivations.

MLP Team Structure

The Design-Forward MLP Team (4-6 people):

- 1 Product Designer (UX research, strategy, emotional design)
- 1 Visual/Interaction Designer (interface design, motion, delight)
- 2-3 Engineers (implementation, performance, reliability)
- 1 User Researcher (emotional testing, feedback analysis)

MLP teams are design-heavy compared to traditional development teams. Design skills and sensibilities are critical for success.

MLP Budget Allocation

40% Design/UX: User research, interface design, interaction design, emotional testing

40% Engineering: Technical implementation optimized for user experience

15% User Research: Emotional validation, sentiment analysis, user testing

5% Other: Tools, infrastructure, coordination

MLP budgets invest more heavily in design than any other approach, reflecting the priority placed on emotional engagement.

MLP Timeline Planning

Week 1-2: Emotional user research and design strategy

Week 3-4: Initial design concepts and emotional testing

Week 5-8: Iterative design and development with continuous emotional validation

Week 9-10: Polish and delight feature implementation

Week 11-12: Final emotional testing and launch preparation

MLP timelines include multiple rounds of emotional testing and design refinement that other approaches skip.

MLP Resource Requirements

Design Talent:
Access to designers who understand emotional design principles, not just functional design.

User Testing Infrastructure:
Ability to conduct frequent emotional testing with real users.

Development Flexibility:
Engineering processes that can accommodate design-driven changes.

Quality Expectations:
Users who will appreciate and respond to thoughtful design details.

MMP Resource Allocation: Sales-Focused Investment for Revenue Generation

MMP resource allocation optimizes for features and capabilities that drive purchasing decisions, often requiring go-to-market resources that other approaches don't need.

The MMP Resource Philosophy

Revenue-Driven Prioritization:
Features are prioritized based on their impact on sales conversion and deal size.

Professional Polish Requirements:
Building for buyers requires higher quality standards than building for users.

Sales Enablement Investment:
Resources dedicated to sales materials, demos, and buyer education.

Market Readiness Focus:
Ensuring the product can compete effectively against established alternatives.

MMP Team Structure

The Market-Ready MMP Team (6-8 people):

- 1 Product Manager (buyer research, competitive analysis, requirements)

- 2-3 Engineers (feature development, integration, reliability)
- 1 Designer (professional interface design, sales materials)
- 1 Sales/Marketing person (buyer research, positioning, sales enablement)
- 1 QA/Support person (quality assurance, customer success)

MMP teams include go-to-market roles that other approaches don't require, reflecting the focus on sales readiness.

MMP Budget Allocation

45% Engineering: Feature development focused on buyer requirements

20% Sales/Marketing: Buyer research, positioning, sales materials

15% Design/UX: Professional interface design and user experience

10% Quality Assurance: Reliability and professional quality standards

10% Other: Infrastructure, tools, competitive analysis

MMP budgets include significant sales and marketing investment to ensure market readiness.

MMP Timeline Planning

Week 1-2: Buyer research and competitive analysis

Week 3-4: Requirements definition based on purchase criteria

Week 5-10: Feature development with continuous buyer validation

Week 11-12: Sales enablement and professional polish

Week 13-14: Market launch with sales support

MMP timelines include buyer research and sales enablement activities that pure product approaches don't require.

MMP Resource Success Factors

Buyer Access:
Direct access to people who make purchasing decisions in your target market.

Sales Skills:
Team members who understand buyer psychology and sales processes.

Competitive Intelligence:
Resources to understand and respond to competitive alternatives.

Professional Standards:
Ability to deliver quality that meets buyer expectations for paid products.

The Resource Investment Curves by Approach

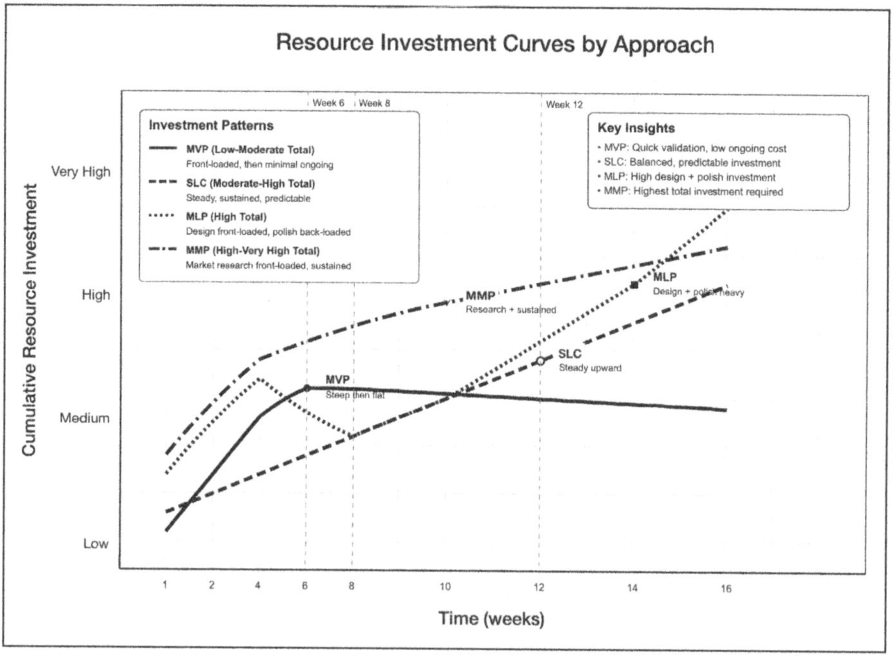

Investment Patterns Shown:

1. **MVP Curve**:
   Very steep investment weeks 1-6, sharp plateau weeks 7-8, then relatively flat ongoing investment.
   Total Investment: Low to moderate.
   Investment Pattern: Front-loaded, then minimal ongoing

2. **SLC Curve**:
   Moderate initial investment, steady consistent slope throughout all 16 weeks.
   Total Investment: Moderate to high.
   Investment Pattern: Steady, sustained, predictable

3. **MLP Curve**:
   High initial design investment weeks 1-4, moderate middle phase, very high polish investment weeks 13-16.
   Total Investment: High.
   Investment Pattern: Design-front-loaded, polish-back-loaded

4. **MMP Curve**:
   High market research investment weeks 1-4, sustained high investment throughout development and sales enablement.
   Total Investment: High to Very High.
   Investment Pattern: Market research front-loaded, sustained high investment

The diagram clearly illustrates how different product approaches require fundamentally different resource allocation strategies, with implications for budgeting, team planning, and timeline expectations. The phase markers at weeks 6, 8, and 12 highlight critical transition points where investment patterns shift for each approach.

Key Insights from the Curves:

- **MVP** has the lowest total investment but requires discipline to avoid scope creep
- **SLC** has the most predictable investment pattern and timeline
- **MLP** requires design expertise and has unpredictable polish requirements
- **MMP** has the highest total investment due to market readiness requirements

The curves show why choosing the wrong approach for your resource constraints leads to project failure or massive scope reduction.

Team Skill Requirements by Approach

Different approaches require different skill mixes within your team. Hiring for the wrong approach wastes resources and delays progress.

MVP Team Skills

Essential Skills:

- Rapid prototyping and development
- Technical decision-making under uncertainty
- User research and testing
- Scope management and ruthless prioritization

Nice-to-Have Skills:

- Visual design
- Marketing and positioning
- Customer success and support

Critical Characteristics:

- Comfort with ambiguity and rapid change
- Willingness to build imperfect solutions
- Strong communication for small team coordination

SLC Team Skills

Essential Skills:

- Full-stack development capabilities
- User experience design and research
- Quality assurance and testing
- Project management and coordination

Nice-to-Have Skills:

- Specialized domain expertise
- Marketing and go-to-market
- Customer success and community building

Critical Characteristics:

- Collaborative work style across disciplines
- Quality orientation without perfectionism
- Systematic thinking about complete user workflows

MLP Team Skills

Essential Skills:

- Emotional design and user psychology
- Interaction design and visual design
- User research focused on emotional response
- Development with attention to user experience details

Nice-to-Have Skills:

- Community building and social dynamics
- Content creation and storytelling
- Brand strategy and positioning

Critical Characteristics:

- Empathy and emotional intelligence
- Attention to detail and craft
- Understanding of human psychology and motivation

MMP Team Skills

Essential Skills:

- Buyer research and competitive analysis
- Sales and marketing strategy
- Professional quality development and design
- Customer success and support

Nice-to-Have Skills:

- Industry domain expertise
- Enterprise sales experience
- Regulatory and compliance knowledge

Critical Characteristics:

- Customer-facing communication skills
- Professional quality standards
- Understanding of buyer psychology and decision-making

Budget Planning by Approach

Understanding the financial implications of each approach helps you plan realistically and avoid budget surprises.

MVP Budget Planning

Typical MVP Budget Range: $10,000 - $75,000

Primary Cost Drivers: Engineering time, basic tools and infrastructure

Hidden Costs: Technical debt cleanup, scope creep, team scaling after validation

Budget Breakdown Example (8-week MVP, 3-person team):

- Engineering: $35,000 (70%)
- Design: $10,000 (20%)
- Tools/Infrastructure: $2,500 (5%)
- User Testing: $2,500 (5%)
- **Total**: $50,000

SLC Budget Planning

Typical SLC Budget Range: $75,000 - $250,000

Primary Cost Drivers: Balanced team costs, longer timeline, quality requirements

Hidden Costs: Integration complexity, user research, iterative refinement

Budget Breakdown Example (14-week SLC, 6-person team):

- Engineering: $70,000 (50%)
- Design/UX: $42,000 (30%)
- Quality Assurance: $14,000 (10%)
- Tools/Infrastructure: $7,000 (5%)
- User Research: $7,000 (5%)
- **Total**: $140,000

MLP Budget Planning

Typical MLP Budget Range: $100,000 - $300,000

Primary Cost Drivers: Design expertise, emotional testing, polish iterations

Hidden Costs: Multiple design iterations, specialized design talent, emotional testing

Budget Breakdown Example (12-week MLP, 5-person team):

- Design/UX: $60,000 (40%)
- Engineering: $60,000 (40%)
- User Research: $22,500 (15%)
- Tools/Infrastructure: $7,500 (5%)
- **Total**: $150,000

MMP Budget Planning

Typical MMP Budget Range: $150,000 - $500,000

Primary Cost Drivers: Market research, professional quality, sales enablement

Hidden Costs: Competitive analysis, buyer research, sales material creation

Budget Breakdown Example (14-week MMP, 7-person team):

- Engineering: $81,000 (45%)
- Sales/Marketing: $36,000 (20%)
- Design/UX: $27,000 (15%)
- Quality Assurance: $18,000 (10%)
- Tools/Infrastructure: $18,000 (10%)
- **Total**: $180,000

Risk Management Across Approaches

Different approaches carry different types of risks that require different mitigation strategies.

MVP Risk Profile

Primary Risks:

- Technical debt accumulation
- User disappointment from incomplete experience
- Difficulty scaling team and technology after validation
- Learning the wrong lessons from limited functionality

Risk Mitigation Strategies:

- Set clear technical debt repayment plans
- Manage user expectations about product completeness
- Plan for post-MVP development architecture
- Focus testing on core assumptions rather than edge cases

SLC Risk Profile

Primary Risks:

- Scope creep leading to timeline delays
- Over-engineering for current needs
- Competing priorities between simplicity and completeness
- Market changes during longer development cycles

Risk Mitigation Strategies:

- Rigorous scope management and change control
- Regular user testing to validate decisions
- Clear definition of "complete" within chosen scope
- Market monitoring and competitive analysis

MLP Risk Profile

Primary Risks:

- Subjective design decisions causing delays
- Difficulty measuring emotional success
- Design-engineering coordination challenges
- Market not valuing emotional differentiation

Risk Mitigation Strategies:

- Establish emotional design criteria upfront
- Implement regular emotional testing and validation
- Create collaborative design-engineering processes
- Validate emotional value proposition before heavy investment

MMP Risk Profile

Primary Risks:

- Overbuilding for buyer requirements vs. user needs
- Competitive response during development
- Sales cycle longer than expected
- Professional quality standards causing delays

Risk Mitigation Strategies:

- Balance buyer and user requirements throughout development
- Monitor competitive landscape and adjust positioning
- Plan for longer sales cycles in budget and timeline
- Define quality standards early and stick to them

Timeline Planning and Milestone Management

Effective resource allocation requires realistic timeline planning that accounts for the specific characteristics of each approach.

Timeline Planning Principles

Buffer for Unknowns:
Each approach has predictable areas of uncertainty that require timeline buffers.

Milestone-Based Progress:
Break development into measurable milestones that enable course correction.

Resource Ramp Planning:
Account for team ramp-up time and skill development.

Market Timing Considerations:
Align development timelines with market opportunities and constraints.

MVP Timeline Characteristics

Shortest Overall Timeline: 4-8 weeks typical

Highest Uncertainty: Technical discoveries can dramatically impact timeline

Scope Flexibility: Timeline is fixed, scope adjusts to fit

Quick Iteration Cycles: 1-2 week iterations with regular user feedback

SLC Timeline Characteristics

Moderate Timeline: 12-16 weeks typical

Predictable Progress: Balanced approach enables better timeline estimation

Quality Gates: Regular quality checkpoints that can extend timeline

User Feedback Integration: Built-in time for user testing and iteration

MLP Timeline Characteristics

Variable Timeline: 8-16 weeks depending on emotional validation cycles

Design Iteration Uncertainty: Emotional testing can require multiple design iterations

Polish Buffer Required: Final polish phase can extend timeline significantly

Emotional Validation Gates: Progress blocked until emotional goals are met

MMP Timeline Characteristics

Longest Timeline: 12-20 weeks typical

Market Research Frontend: Significant upfront time for buyer and competitive research

Professional Quality Requirements: Quality standards can extend timeline

Sales Enablement Backend: Additional time for sales materials and training

Resource Optimization Strategies

Once you understand the resource requirements of your chosen approach, these strategies can help you optimize efficiency without compromising outcomes.

Cross-Functional Skill Development

MVP Teams:
Engineers learn basic design, designers learn technical constraints

SLC Teams:
All team members understand complete user workflows

MLP Teams:
Engineers develop aesthetic sense, designers understand technical possibilities

MMP Teams:
Product people develop sales skills, engineers understand buyer requirements

Outsourcing and Partnership Strategies

What to Outsource:

- Non-core capabilities that would take too long to develop internally
- Specialized skills needed for short periods
- Well-defined work that doesn't require deep domain knowledge

What to Keep Internal:

- Core product decisions and strategy
- Direct user interaction and feedback
- Skills that will be needed long-term
- Work that requires rapid iteration and change

Tool and Infrastructure Investment

MVP Tools:
Prioritize speed over sophistication

SLC Tools:
Balance capability with ease of use

MLP Tools:
Emphasize design and collaboration capabilities

MMP Tools:
Include sales and customer management capabilities

Your Resource Planning Action Plan

Week 1: Resource Assessment

- Audit your current team skills against your chosen approach
- Calculate realistic budget requirements for your approach
- Identify skill gaps and resource constraints
- Plan team development or hiring strategy

Week 2: Timeline Planning

- Create realistic timeline based on approach requirements
- Identify major milestones and risk points
- Plan buffer time for approach-specific uncertainties
- Align timeline with market opportunities

Week 3: Risk Planning

- Identify primary risks for your chosen approach

- Develop mitigation strategies for each major risk
- Create contingency plans for timeline or budget overruns
- Establish early warning systems for major risks

Week 4: Resource Optimization

- Implement cross-functional skill development plans
- Evaluate outsourcing opportunities for non-core work
- Invest in tools and infrastructure appropriate to your approach
- Create resource monitoring and adjustment processes

Remember: Your resource constraints are real, but they don't have to be permanent. The key is choosing an approach that fits your current constraints while building capabilities for your future ambitions.

In the next chapter, we'll explore how to manage the journey from launch to product-market fit, including how to evolve your approach as your product and market mature.

Your resources are finite, but your potential isn't. Plan wisely, execute efficiently, and build the foundation for long-term success.

7

From Launch to Product-Market Fit

I have a theory about product launches that might sound counterintuitive: the launch is the least important day in your product's life. I've watched founders spend months perfecting launch strategies, coordinating press releases, and timing social media campaigns, only to discover that nobody cares about their "big reveal." Meanwhile, there are products that often succeed by launching quietly, iterating furiously, and finding product-market fit through dozens of small improvements rather than one perfect debut.

This chapter is about what happens after the confetti settles and the real work begins. Because regardless of whether you built an MVP, SLC, MLP, or MMP, your path to product-market fit will be paved with user feedback, competitive pressure, and constant decisions about what to build next. The question isn't whether you'll need to evolve your product; it's whether you'll do so intelligently.

The Post-Launch Reality Check

Here's what every founder discovers after launch: your product development approach doesn't end when you ship. It evolves based on market feedback, competitive dynamics, and resource constraints. The MVP you launched might need to become more lovable. The MLP you built might need to become more marketable. Success requires adapting your approach based on what you learn from real users in real markets.

The Three Post-Launch Phases

Phase 1: Initial Market Response (Weeks 1-4)
Your first users interact with your product and form initial opinions. Their behavior tells you whether your chosen approach matches market expectations.

Phase 2: Pattern Recognition (Months 2-6)
You start seeing patterns in user behavior, feedback, and market response. These patterns guide your evolution strategy.

Phase 3: Approach Evolution (Months 6+)
You deliberately evolve your approach based on what you've learned, often transitioning from one methodology to another.

Understanding these phases helps you plan for the inevitable evolution rather than being surprised by it.

Common Post-Launch Discoveries

The "Good Enough" Gap:
Your MVP works fine, but users keep asking, "When will you add [obvious feature]?" You realize minimal isn't viable in your market.

The "Love But Won't Pay" Problem:
Users adore your MLP but won't upgrade to paid plans. Emotional engagement doesn't automatically translate to revenue.

The "Feature Fatigue" Syndrome:
Your SLC product is complete, but users only use 20% of the features regularly. Completeness might be overwhelming rather than helpful.

The "Buyers vs. Users" Disconnect:
Your MMP converts buyers well, but daily users struggle with complexity designed for purchase criteria rather than usage patterns.

These discoveries aren't failures; they're learning opportunities that guide your evolution strategy.

Understanding User Feedback Across Different Approaches

Different product approaches generate different types of feedback, and interpreting that feedback correctly is crucial for intelligent evolution.

MVP Feedback Interpretation

What MVP Users Typically Say:

- "It's a good start, but..."
- "I wish it could also..."
- "It would be perfect if you added..."
- "I'll wait until you build more features"

What This Feedback Means:

- Your core assumption is validated (good start)
- You've identified the next most important features (wish list)
- Users can envision using a more complete version (positive signal)
- Your current version isn't complete enough for regular use (scope guidance)

How to Respond:
Don't immediately build every requested feature. Instead, look for patterns in feedback that point toward the most critical completeness gaps. Prioritize features that move users from "interesting" to "essential."

SLC Feedback Interpretation

What SLC Users Typically Say:

- "This is exactly what I needed"
- "It just works the way I expected"
- "I can finally get my job done in one place"
- "Why would I use anything else?"

What This Feedback Means:

- You've achieved product-market fit for your chosen scope (validation)
- Users appreciate the complete workflow approach (differentiation confirmed)
- You're creating switching costs through integration (defensibility)
- Market expansion opportunities may exist (growth potential)

How to Respond:
Focus on depth over breadth. Make your core workflow even better before expanding scope. Look for adjacent use cases that extend your completeness advantage.

MLP Feedback Interpretation

What MLP Users Typically Say:

- "I love using this"
- "It makes me feel [positive emotion]"
- "I recommended it to all my friends"
- "I hope you don't change it too much"

What This Feedback Means:

- You've created emotional attachment (differentiation achieved)
- Word-of-mouth growth potential is high (scalable acquisition)
- User retention should be strong (economic validation)
- Changes need to preserve emotional connection (evolution constraint)

How to Respond:
Protect the emotional core while expanding utility. Add features that enhance the love rather than diluting it. Consider premium offerings that monetize the emotional attachment.

MMP Feedback Interpretation

What MMP Users Typically Say:

- "This is worth the price"
- "It's better than [competitor]"
- "It solved our problem immediately"
- "We're seeing ROI already"

What This Feedback Means:

- Value proposition is clear and validated (pricing confirmed)
- Competitive positioning is working (market position secured)
- Customer success is measurable (retention predicted)
- Expansion revenue opportunities exist (growth path identified)

How to Respond:
Focus on customer success and expanding revenue. Build features that increase deal sizes and reduce churn. Document and replicate your successful customer patterns.

Scaling Decisions: Maintaining Quality vs. Adding Features

Every successful product faces the same tension: users want more features, but adding features can dilute the core experience that made the product successful in the first place. How you navigate this tension depends on your original approach and market position.

The Feature Addition Framework

Before adding any feature, regardless of your approach, ask these questions:

Does this feature strengthen our core differentiation or dilute it?

- Strengthening features should get priority.
- Diluting features should be rejected or reconsidered.

Does this feature serve our existing users better or attract new users?

- Existing user features typically have higher ROI.
- New user features carry a higher risk but broader potential.

Does this feature require new infrastructure or leverage existing capabilities?

- Leveraging existing capabilities is faster and cheaper.
- New infrastructure should serve multiple future features.

Does this feature align with our long-term vision or solve a short-term problem?

- Vision-aligned features build toward something bigger.
- Short-term fixes can create technical debt.

Scaling Strategies by Approach

MVP Scaling Strategy: Selective Completion

MVPs should evolve toward one of the other approaches based on market feedback. The question isn't whether to add features, but which approach to evolve toward.

Evolution Options:

- **Toward SLC**: Add features that complete core workflows.
- **Toward MLP**: Add features that create emotional engagement.
- **Toward MMP**: Add features that justify higher pricing.

SLC Scaling Strategy: Depth Before Breadth

SLC products should resist the temptation to expand scope too quickly. Instead, make your core complete solution even better.

Scaling Priorities:

1. Perfect the core workflow until it's dramatically better than alternatives.
2. Add adjacent workflows that leverage your completeness advantage.
3. Expand to new user segments that need similar complete solutions.

MLP Scaling Strategy: Love-Preserving Enhancement

MLP products must be extremely careful about feature additions that might compromise the emotional experience.

Feature Addition Criteria:

- Does this enhance the emotions we've already created?
- Does this give users new reasons to love the product?
- Does this create more moments of delight rather than just utility?

MMP Scaling Strategy: Value Stack Expansion

MMP products should focus on features that increase deal sizes and customer lifetime value.

Revenue-Focused Features:

- Enterprise features that justify higher pricing tiers.
- Integration capabilities that increase switching costs.
- Analytics and reporting that demonstrate ROI.
- Multi-user features that expand within accounts.

Competitive Response and Market Evolution

Markets don't stay static after you launch. Competitors respond, user expectations evolve, and new technologies emerge. Your post-launch strategy needs to account for these changing dynamics.

Competitive Response Patterns

When You Launch an MVP:
Competitors often ignore you initially, assuming you're not a serious threat. This gives you time to evolve toward a more complete offering before they respond.

When You Launch an SLC
Competitors may try to match your completeness with feature additions, but they often struggle to achieve the same integration quality. Your advantage compounds over time.

When You Launch an MLP
Competitors typically try to copy your emotional design elements, but they often miss the underlying psychological insights. Surface-level copying rarely recreates emotional connection.

When You Launch an MMP
Competitors focus on price competition and feature matching. Your advantage lies in market execution and customer success rather than product differentiation alone.

Market Evolution Responses

Emerging to Growing Markets
As markets mature, user expectations increase. MVP approaches become less viable, and SLC or MLP approaches become more important.

Growing to Mature Markets
Competition intensifies, requiring either emotional differentiation (MLP) or clear value propositions (MMP). Pure functionality becomes commoditized.

Technology Disruptions
New technologies can change the entire competitive landscape. Stay close to your users' underlying jobs rather than getting attached to specific solutions.

Case Study: Zoom's Evolution to Market Leadership

Zoom entered the video conferencing market as a classic MMP, focusing on building something that businesses would actually pay for immediately, rather than just tolerate and use occasionally. Their evolution from launch to market dominance illustrates how intelligent post-launch strategy can overcome entrenched competitors.

Zoom's Initial MMP Strategy

When Zoom launched in 2013, the video conferencing market was dominated by Skype (consumer-focused) and WebEx (enterprise-focused). Both had significant market share and brand recognition. Zoom's MMP approach focused on what businesses actually needed from video conferencing, not just what existing products offered.

Marketable from Day One:

- Reliable, consistent HD video quality that worked consistently
- Simple one-click meeting setup that didn't require technical expertise
- Competitive pricing that justified switching from alternatives: $15/user/mo - transparent value
- Cross-platform compatibility that solved real business problems

Revenue-Focused Features:

- Business admin controls that enterprise buyers required
- Integration capabilities with existing business tools
- Recording and transcription features that added measurable value
- Scalable pricing that grew with customer needs

Zoom's Post-Launch Evolution

Rather than staying purely MMP, Zoom evolved their approach based on market feedback and competitive dynamics:

Phase 1: MMP to SLC Evolution (2013-2016)
Zoom realized that businesses needed complete meeting solutions, not just video calling. They expanded to handle the entire meeting workflow:

- Pre-meeting: Scheduling integration, invitation management
- During meeting: Screen sharing, recording, breakout rooms (2015)
- Post-meeting: Recording distribution, follow-up integration

This SLC evolution made Zoom a complete meeting software solution rather than just a video calling tool.

Phase 2: SLC to MLP Elements (2016-2020)
As competition intensified, Zoom added emotional elements that made meetings feel more human and enjoyable:

- Virtual backgrounds that reduced meeting anxiety
- Reaction features that made meetings more interactive
- Waiting room features that gave hosts control and confidence
- Gallery view that made large meetings feel more inclusive

These MLP elements didn't compromise the business focus but made the experience more human and angaging.

Phase 3: Market Leadership Through Completeness (2020+)
When COVID-19 suddenly made video conferencing essential for everyone, Zoom's complete, lovable, marketable solution scaled to meet unprecedented demand, handling 300M daily users flawlessly. Their evolution strategy positioned them perfectly for market expansion.

Key Evolution Lessons from Zoom

Start with a Clear Market Focus:
Zoom's initial MMP focus on business buyers created a strong foundation for evolution.

Evolve Based on User Jobs:
They expanded by understanding the complete job of "running effective meetings," not just "video calling."

Preserve Core Strengths:
Throughout evolution, Zoom has maintained its advantages in reliability and simplicity.

Add Emotional Elements Carefully:
MLP features enhanced rather than replaced their business value proposition.

Case Study: Figma's Community-Driven Growth

Figma's evolution illustrates how MLP approaches can scale through community engagement and network effects, creating sustainable competitive advantages that are difficult for incumbents to replicate.

Figma's Initial MLP Strategy

When Figma launched in 2016, Adobe Creative Suite dominated design tools. Rather than competing on features alone, Figma built an MLP that made designers feel more creative and collaborative.

Lovable from the Start:

- Browser-based design that felt magical in 2016

- Real-time collaboration that made design feel social
- Component systems that made designers feel more systematic
- Performance that made design work feel fluid rather than frustrating

Emotional Identity Reinforcement: Figma made designers feel like they were part of a modern, forward-thinking design community rather than just users of traditional tools.

Figma's Community-Driven Evolution

Figma's post-launch strategy focused on amplifying the community and collaborative aspects that made their MLP successful:

Phase 1: Community Platform Development (2016-2018)
Instead of just adding features, Figma built capabilities and transparency that strengthened designer communities:

- File sharing that made design work more transparent
- Component libraries that enabled design system collaboration
- Version control that made team design work more reliable
- Commenting systems that improved design feedback processes

Phase 2: Network Effect Amplification (2018-2020)
Figma realized that their collaborative advantage created network effects where each new user made the platform more valuable for existing users:

- Team features that made adoption viral within organizations
- Educational content that built design community knowledge
- Plugin ecosystem that let the community extend Figma's capabilities
- Public file galleries that showcased community creativity

Phase 3: Market Expansion Through Community (2020-2022)
Figma's strong community became a competitive moat that enabled expansion into adjacent markets:

- FigJam for broader collaboration beyond design
- Dev Mode for designer-developer collaboration
- Advanced prototyping for design-to-product workflows
- Enterprise features that leveraged existing community adoption

Key Evolution Lessons from Figma

Emotional Connection Scales:
Figma's focus on making designers feel good about their work created sustainable competitive advantages.

Community as Moat:
Building genuine community created network effects that were difficult for competitors to replicate.

Collaborative Advantage Compounds:
Features that made collaboration better created exponential value rather than linear improvements.

Expand Through Community Needs:
Figma's expansion strategy followed community needs rather than arbitrary feature additions.

Transition Strategies: Moving Between Approaches

Sometimes your initial approach proves to be wrong for your market, and you need to transition to a different methodology. These transitions are challenging but often necessary for long-term success.

Common Transition Patterns

MVP to SLC Transition When to Consider:
User feedback consistently requests complete workflows rather than individual features. **How to Execute:** Map complete user jobs and prioritize features that enable end-to-end task completion. **Resources Required:** Additional design and integration work to create cohesive experiences.

MVP to MLP Transition When to Consider:
Users like your functionality but aren't emotionally engaged or advocating for your product. **How to Execute:** Invest in user research to understand emotional needs and redesign experiences to create delight. **Resources Required:** Design expertise and emotional testing capabilities.

SLC to MLP Transition When to Consider:
Your complete solution works well but doesn't create strong user preference or loyalty. **How to Execute:** Add emotional elements that enhance rather than replace functional completeness. **Resources Required:** Emotional design skills and user psychology understanding.

MLP to MMP Transition When to Consider:
Users love your product but aren't willing to pay prices that support your business model. **How to Execute:** Add features and positioning that appeal to buyers rather than just users. **Resources Required:** Sales and marketing expertise, buyer research capabilities.

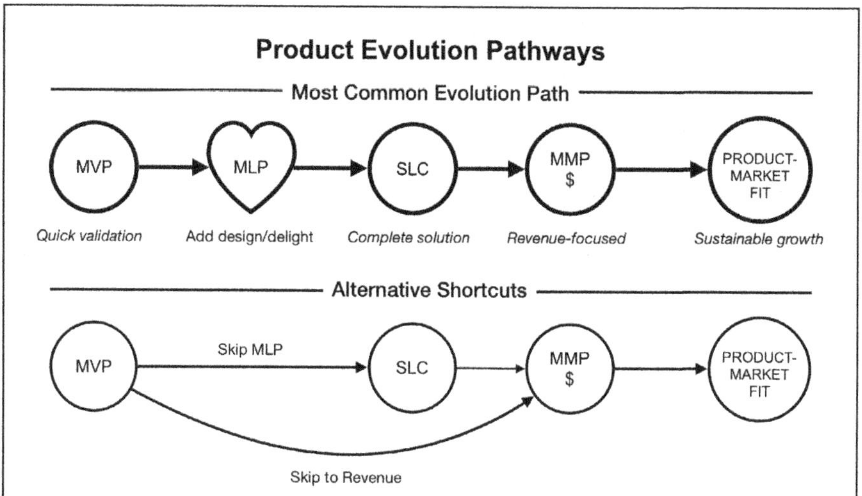

Evolution Stages Summary

Stage	Purpose	Key Focus	Typical Outcome
MVP	Quick validation	Test core hypothesis	Early feedback
MLP	Add design/delight	User engagement	Emotional connection
SLC	Complete solution	Full functionality	Broader adoption
MMP	Revenue-focused	Monetization	Sustainable revenue

Key Insights:

- Most products follow MVP → MLP → SLC → MMP progression
- Shortcuts are possible but less common

Transition Planning Framework

Phase 1: Transition Decision (Week 1-2)

- Analyze current approach performance against business goals
- Identify specific gaps that alternative approaches might address
- Evaluate resource requirements for potential transitions
- Choose target approach based on market evidence and capabilities

Phase 2: Transition Strategy (Week 3-4)

- Map features and capabilities needed for target approach
- Plan resource reallocation and skill development
- Design transition timeline that maintains current user satisfaction
- Identify risk mitigation strategies for transition period

Phase 3: Transition Execution (Month 2-4)

- Implement changes incrementally to avoid disrupting existing users

- Test transition elements with subset of users before full rollout
- Monitor key metrics to ensure transition improves rather than harms performance
- Adjust timeline and scope based on early transition results

Phase 4: Transition Completion (Month 5-6)

- Complete rollout of new approach elements
- Measure success against original transition goals
- Document lessons learned for future evolution decisions
- Plan ongoing development within new approach framework

The Product Evolution Pathways

Core Message: The diagram now clearly communicates that:

- You can start with any approach, the most common and successful path is to begin with MVP for quick validation, then evolve to SLC by adding features and polish. All approaches can ultimately lead to Product-Market Fit, but some paths are more traveled than others.
- There's a logical progression: MVP → MLP → SLC → MMP → Product-Market Fit
- This is the most common and recommended path
- Shortcuts exist but are less common
- All paths lead to the same destination: Product-Market Fit

This simplified approach eliminates the confusion while still showing the strategic options available to product teams.

Key Insights from the Diagram:

- MVP is primarily a starting point that should evolve to other approaches
- SLC and MMP are often stable long-term approaches
- MLP can be either a starting point or an evolution destination
- Most successful products eventually incorporate elements from multiple approaches
- Product-market fit can be achieved through various evolution pathways

Long-Term Product Development Planning

Once you understand your evolution pathway, you need frameworks for making ongoing product decisions that support long-term success.

The Product Development North Star

Your product development decisions should align with a clear vision of where you're headed:

MVP North Star: "Learn our way to product-market fit as quickly as possible"

SLC North Star: "Become the complete solution for [specific job] that users prefer over alternatives"

MLP North Star: "Create the most loved product in our category that users enthusiastically recommend"

MMP North Star: "Build the most valuable solution that buyers choose over alternatives"

This north star guides your decisions about feature prioritization, resource allocation, and evolution decisions.

Feature Prioritization Frameworks

The Approached-Aligned Prioritization Matrix:

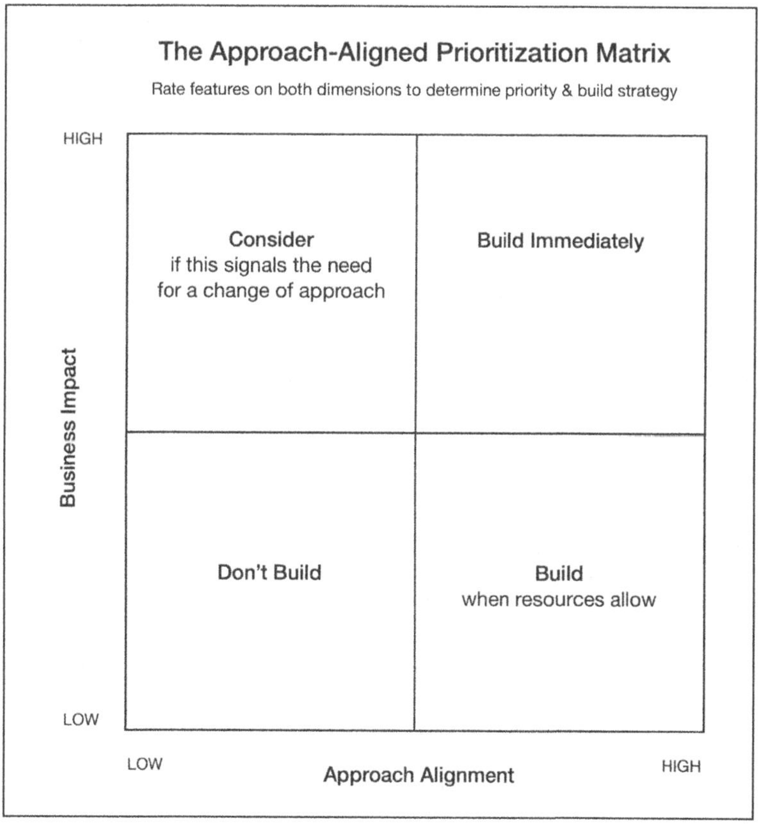

Using the Approach-Aligned Prioritization Matrix (shown on the previous page), rate each potential feature on two dimensions:

- **Approach Alignment:** How well does this feature support our current/ target approach?
- **Business Impact:** How much will this feature contribute to our key business metrics?

Where that feature lands in one of the four quadrants will guide your decision on the action to take.

Success Metrics Evolution

Your success metrics should evolve and change along with your approach:

APPROACH	METRICS	ACTIONS
MVP	**Learning** Metrics	Focus on validated assumptions and iteration velocity rather than just usage
SLC	**Completion** Metrics	Measure end-to-end task success and workflow satisfaction
MLP	**Engagement** Metrics	Track emotional indicators like NPS, retention, and advocacy
MMP	**Revenue** Metrics	Optimize for conversion, deal size, and customer lifetime value

Your Post-Launch Action Plan

Week 1-2: Feedback Analysis

- Collect and categorize all user feedback since launch
- Identify patterns that suggest approach evolution opportunities
- Survey users about satisfaction and feature priorities
- Analyze usage data to understand actual vs. intended product usage

Week 3-4: Evolution Strategy

- Evaluate whether your current approach is achieving business goals
- Consider whether transition to different approach would accelerate success
- Plan resource requirements for potential evolution
- Design evolution timeline that maintains user satisfaction

Months 2-3: Evolution Execution

- Implement highest-priority evolution features

- Test evolution elements with subset of users
- Monitor metrics to ensure evolution improves performance
- Gather feedback on evolution direction from users and stakeholders

Months 4-6: Evolution Assessment

- Measure evolution success against original goals
- Plan ongoing product development within evolved approach
- Document lessons learned for future evolution decisions
- Establish ongoing processes for monitoring evolution opportunities

Remember: Product evolution is an ongoing process, not a one-off choice, involving adapting to market feedback and changing conditions. The companies that succeed in the long run are those that evolve wisely while maintaining their core strengths.

Your product's journey from launch to product-market fit will be unique, but the principles in this chapter will assist you in navigating that journey strategically instead of reactively. The market will tell you what's working and what isn't. Your role is to listen carefully and adapt thoughtfully.

Success isn't about creating the perfect product on your first attempt. It's about making something good enough to start learning, then improving based on that learning until you've created something truly outstanding. Your journey of evolution begins now.

Choosing Your Path Forward

Here's something I often see in startup circles and forums: founders asking, "How long should an MVP take to build?" The answers vary wildly, from "two weeks" to "six months," and they're all wrong. Not because the timelines are incorrect, but because they're answering the wrong question.

The real question isn't "How long should an MVP take?" but rather, "What exactly are you building, and does it align with your definition?" I've seen teams spend six months developing what they label as an MVP, only for it to be a full SLC solution. Conversely, I've watched others launching truly minimal products in markets that expect completeness, and then wonder why users refuse to adopt them.

This disconnect between intention and execution captures why this book exists. The product development approach you choose isn't just a philosophical decision. It influences your timeline, your resource needs, your team setup, and ultimately your chances of success. Getting it wrong doesn't just waste time; it can kill your startup.

But getting it right? That's when the magic happens. When your approach aligns with your market, resources, and goals, product development no longer feels like a constant battle against impossible constraints. It begins to feel like a clear path toward creating something people genuinely want.

The Decision Framework Revisited

Throughout this book, we've explored four distinct approaches to product development, each optimized for different outcomes:

- **MVP**: Optimized for learning speed and assumption validation
- **SLC**: Optimized for complete, delightful user experiences
- **MLP**: Optimized for emotional connection and user love
- **MMP**: Optimized for revenue generation and market readiness

None of these approaches is inherently superior to the others. The right choice depends on your specific situation: your market maturity, customer sophistication, resource constraints, and business goals.

But here's what I've learned after working with hundreds of startups: most founders choose their approach unconsciously, following whatever methodology sounds most appealing or is currently popular in their network. This unconscious choice often results in conscious frustration later when reality doesn't match expectations.

The Choice Points That Matter

As you decide your path forward, these are the critical choice points that will determine your success:

Market Reality vs. Market Aspiration
Build for the market as it exists today, not as you wish it existed. If your market expects complete solutions, don't build an MVP hoping to educate them about the virtues of minimalism.

Resource Constraints vs. Resource Dreams
Plan based on the resources you have, not the resources you hope to raise. A perfectly executed MMP with limited resources beats a half-executed SLC with inadequate funding.

Team Strengths vs. Team Gaps
Choose an approach that leverages your team's existing strengths rather than requiring skills you don't have. An engineering-heavy team will struggle with MLP emotional design. A design-heavy team will struggle with the MMP sales focus.

Short-term Survival vs. Long-term Vision
Balance the need to survive today with the goal of thriving tomorrow. Sometimes you need to start with an approach that generates quick revenue (MMP) even if your long-term vision requires an emotional connection (MLP).

Integration with Customer Discovery

This book builds on the customer discovery foundation established in "Find Problems Worth Solving." That first book helped you identify problems worth solving. This book guides you on how to solve them in ways that create sustainable businesses.

The integration between customer discovery and product development is crucial:

Customer Discovery Informs Approach Selection
The extent of problem validation determines which approaches are viable. Well-validated problems can sustain longer development timelines (SLC), while uncertain problems demand quicker learning cycles (MVP).

Product Approach Enables Customer Success
Choosing the right development approach ensures that your solution creates the customer success that your discovery work promised. Finding real problems is pointless if you solve them poorly.

Ongoing Discovery Guides Product Evolution
Customer discovery doesn't end when you start building. Ongoing customer conversations guide your product evolution from initial approach to market leadership.

The founders who succeed long-term are those who maintain the customer-centric mindset from discovery through development and beyond.

Setting Up for Success in the Next Phase

Regardless of which approach you choose, your success will depend on execution excellence. Here are the principles that separate successful product development from well-intentioned failure:

Principle 1: Clarity Over Consensus

Be crystal clear about which approach you're taking and why. Don't try to please everyone by claiming you're building an "MVP that's also lovable and complete and marketable." Pick one primary approach and optimize for it.

This clarity affects everything from resource allocation to team hiring to success metrics. Ambiguous approaches lead to ambiguous results.

Principle 2: Consistency Over Perfection

Execute your chosen approach consistently rather than perfectly. A well-executed MMP will beat a perfectly conceived but poorly executed SLC every time.

Consistency means making decisions that align with your approach, even when they're difficult or counterintuitive. It also means saying no to features that don't support your main approach goals.

Principle 3: Evolution Over Orthodoxy

Be prepared to evolve your approach based on market feedback. The approach you start with might not be the approach that ultimately wins in your market.

But evolution should be deliberate, not reactive. Change your approach because data suggests a better path, not because you're bored with your current direction.

Principle 4: Customers Over Competitors

Focus on customer success rather than competitive response. Competitors will copy your features, but they can't copy your customer relationships and market understanding.

Build for your customers' success, and competitive advantages will follow naturally. Build primarily to beat competitors, and you'll lose sight of what customers value.

Your Next Steps

As you close this book and return to building your product, here's your path forward:

Week 1: Approach Clarity

- Choose your primary approach based on the frameworks in this book
- Align your team around this choice and its implications
- Adjust your current plans to match your chosen approach
- Set appropriate success metrics for your approach

Week 2: Resource Alignment

- Audit your resources against your chosen approach requirements
- Identify gaps in skills, budget, or timeline
- Make necessary adjustments to the team, scope, or timeline
- Create realistic project plans that match your approach

Week 3: Execution Setup

- Implement appropriate testing and validation methods for your approach
- Establish feedback collection processes that match your methodology
- Create decision-making frameworks aligned with your approach
- Begin building with clear approach guidelines

Week 4: Ongoing Learning

- Establish regular check-ins to assess approach effectiveness
- Plan for potential evolution based on market feedback
- Document lessons learned for future product decisions
- Share insights with your team and stakeholders

Beyond Month 1: Mastery Through Practice

The concepts in this book become valuable only through application. Every product decision you make is an opportunity to practice approach-specific thinking. Every user interaction is a chance to validate whether your chosen approach matches market reality.

Preview of What's Next

This book focused on building products that people want. The next book in "The Startup Success Series" will focus on building businesses that can scale and thrive long-term. We'll explore how different product approaches affect your business model, go-to-market strategy, and organizational development.

But that's future learning. Your current opportunity is to apply these frameworks to your product development challenges today.

The Bottom Line Up Front

After everything we've covered, here's what matters most:

There is no universally "right" product development approach.
The best approach is the one that aligns with your market, resources, and goals while setting you up for long-term success.

Most product failures stem from approach mismatch, not execution failure.
Teams that pick the wrong approach and carry it out perfectly still fail. Teams that choose the right approach and execute it properly often succeed.

Your first approach choice doesn't have to be your last.
Successful products often develop through different approaches as they scale and markets mature. Plan for this evolution, but execute your current strategy with focus and discipline.

Customer success is the ultimate validation.
No matter which approach you choose, your success is determined by whether customers reach their goals with your product. Everything else is secondary.

Your Competitive Advantage

In a world where anyone can create a product over a weekend, your competitive edge doesn't come from having better tools or faster development. It comes from making smarter decisions about what to build and how to build it.

The frameworks in this book give you that decision-making edge. While your competitors are unknowingly sticking to the latest popular methodology, you can deliberately choose the approach that boosts your chances of success.

While they're building MVPs because someone told them to "fail fast," you can choose an approach that genuinely fits your market and resources.

While they're adding features randomly based on the loudest user feedback, you can develop your product strategically using approach-specific insights.

This strategic thinking advantage compounds over time. Every good decision makes the next one easier. Every aligned choice creates momentum towards your goals.

The Final Challenge

I'll leave you with the same challenge I give every founder I work with:

For the next 90 days, make every product decision through the lens of your selected approach.

When users request features, ask: "Does this align with our approach?"

When competitors launch new capabilities, ask: "How does this affect our approach strategy?"

When you face resource constraints, ask: "What would our approach prioritize?"

When you celebrate successes or analyze failures, ask: "What does this teach us about our approach?"

This disciplined approach will turn your product development from a series of reactive choices into a clear strategy that leads to long-term success.

Now Go Build

You have the frameworks. You understand the trade-offs. You know the success patterns and failure modes. The only thing left is to build something remarkable.

Your customers are waiting for a product that not only solves their problems but does so in a way that makes them feel successful, capable, and proud of their decision.

Your market is ready for a solution that cuts through the noise of minimal, barely-viable alternatives.

Your opportunity is to select the right approach and implement it so effectively that your product becomes the clear choice for your target customers.

The difference between startups that succeed and those that fail often comes down to this: successful founders make deliberate choices about how to build their products, while unsuccessful founders let their approach choose them.

You now know to be deliberate. You understand the implications of each choice. You can see the path from your current position to achieving product-market fit.

The question isn't whether you can build something people want, but whether you'll choose the right way to build it.

Your customers are waiting. Your market is ready. Your competitive advantage is within reach.

Now go create something worth choosing.

Made in the USA
Monee, IL
04 November 2026

67380409R00075